AF621554

Best Practice Guide on the Control of Iron and Manganese in Water Supply

Metals and Related Substances in Drinking Water Series

Best Practice Guide on the Control of Iron and Manganese in Water Supply

Edited by

Dr Adam Postawa and Dr Colin Hayes

Publishing

London • New York

Published by **IWA Publishing**
Alliance House
12 Caxton Street
London SW1H 0QS, UK
Telephone: +44 (0)20 7654 5500
Fax: +44 (0)20 7654 5555
Email: publications@iwap.co.uk
Web: www.iwapublishing.com

First published 2013

British Library Cataloguing in Publication Data
A CIP catalogue record for this book is available from the British Library

ISBN: 9781780400044 (Paperback)
ISBN: 9781780400747(eBook)

Contents

About this Best Practice Guide

This Guide is one of a series produced by the International Water Association's Specialist Group on Metals and Related Substances in Drinking Water. It is a state-of-the-art compilation of the range of regulatory, scientific and operational issues concerned with the control of iron and manganese in drinking water. It will be of interest to water utility practitioners, researchers, health agencies and policy makers.

The Specialist Group (and its predecessor COST Action 637) is an active research network and has regularly convened international conferences and seminars. It has close working links with the World Health Organization, the European Commission's Joint Research Centre, Health Canada and the US Environmental Protection Agency.

Authors

Matthew Bower, Drinking Water Quality Regulator for Scotland (UK)
Dr Brian Croll, WQM Associates Ltd (UK)
Prof Margherita Ferrante, Catania University (IT)
Prof Józef Górski, Adam Mickiewicz University (PL)
Dr Colin Hayes, WQM Associates Ltd and Swansea University (UK)
Dr Joanna Jeż-Walkowiak, Poznań University of Technology (PL)
Dr Dragana Jovanovic, Institute of Public Health of Serbia (RS)
Martin Jung, AIT Austrian Institute of Technology GmbH (AT)
Dr Michał Michałkiewicz, Poznań University of Technology (PL)
Dr Adam Postawa, AGH University of Science and Technology (PL)
Dr Alina Pruss, Poznan University of Technology (PL)
Dr Ingegerd Rosborg, Consultant (SE)
Dr Stanisław Rybicki, Cracow University of Technology (PL)
Dr Marcin Siepak, Adam Mickiewicz University (PL)
Prof Marek M. Sozański, Poznań University of Technology (PL)
Dr Agnieszka Szuster–Janiaczyk, Poznań University of Technology (PL)

Acknowledgements

This Best Practice Guide derives from the knowledge initially assembled by the European research network COST Action 637, supported by a wide range of experts from 27 European countries, the US and Canada. The funding received from COST over the period December 2006 to November 2010 is duly acknowledged.

The main objective of COST Action 637 was to stimulate better control of metals in drinking water and to minimise environmental and health impacts. The Action aimed to contribute to the implementation of the European Drinking Water Directive and to the Environment and Health programme of the EU Commission and the UN/WHO Protocol on Water and Health. The Joint Research Centre of the European Commission provided scientific input in the COST Action but was also the link to the European Commission. COST stands for European Cooperation in Science and Technology and is the oldest and widest European inter-governmental network for cooperation in research. COST is supported by the EU RTD Framework programme.

Since November 2010, the research network has continued to be active as a Specialist Group within the International Water Association.

Thanks are due to all the water companies, regulators, health agencies and research institutes who have provided access to data.

Review panel

The authors and publishers wish to thank all those involved in the review of this Best Practice Guide. The review involved the following:

Dr Brian Croll, WQM Associates Ltd (UK)
Matthew Bower, Drinking Water Quality Regulator, Scotland (UK)
Dr Darren Lytle, US Environmental Protection Agency (US)

Acronyms

30MS	30 Minutes Stagnation Sampling
AAS	Atomic Absorption Spectroscopy
AI	Adequate Intake
ACC	American Chemistry Council
ASTM	American Society for Testing And Materials
AWWA	American Water Works Association
AWWARF	American Water Works Association Research Foundation
CIPP	Cured-in-Place-Pipe
DAC	Daily Average Concentration
DIC	Dissolved Inorganic Carbon
DO	Dissolved Oxygen
EU	European Union
F AAS	Flame Atomic Absorption Spectroscopy
GF AAS	Graphite Furnace Atomic Absorption Spectroscopy
GIS	Geographic Information System
GLM	Generalised Linear Modelling
HA	Health Advisories
HDPE	High-density Polyethylene
HPLC	High Performance Liquid Chromatography
HRT	Hydraulic Retention Time
ICP-AES	Inductively Coupled Plasma Atomic Emission Spectrometry
ICP-MS	Inductively Coupled Plasma Mass Spectrometry
ICP-OES	Inductively Coupled Plasma Optical Emission Spectrometry
JECFA	Joint (FAO/WHO) Expert Committee on Food Additives
IR	Iron Removal
IR/MR	Iron and Manganese Removal
ISO	International Organization for Standardization

JECFA	Joint Expert Committee on Food Additives
MCL	Maximum Contaminant Level
MIC	Microbially Influenced Corrosion
MR	Manganese Removal
NPDWR	National Primary Drinking Water Regulations (US)
NSDWR	National Secondary Drinking Water Regulations (US)
PD	Parkinson's Disease
PMTDI	Provisional Maximum Tolerable Daily Intake
RDA	Recommended Dietary Allowance
RDT	Random Daytime Sampling
RfD	Reference Dose
RO	Reverse Osmosis
SDWA	Safe Drinking Water Act (US)
SMCL	Secondary Maximum Contaminant Level
TDI	Tolerable Daily Intake
UK	United Kingdom
UN	United Nations
US	United States
US EPA	US Environmental Protection Agency
WHO	World Health Organization
WTP	Water Treatment Plant

Definitions

Tuberculation	Development or formation of small mounds of corrosion products on the inside of iron or steel pipe. Tubercles increase the roughness of the inside of the pipe thus increasing resistance to water flow.
Compliance monitoring	Monitoring at the point of compliance to verify that water supplied for human consumption is in compliance with its quality requirements.
First draw sample	Sample that is taken first in the morning, before the tap in the premise has been used for other purposes.
Fully flushed sample	Sample that is taken after prolonged flushing of the tap
Inventory monitoring	Monitoring to estimate the metal problems in a water supply zone or country.
Operational monitoring	Monitoring activities to check the quality of source waters, to validate the operation of the water treatment processes, or to confirm water quality within distribution networks and domestic distribution systems up to the point of use.
Proportional sample	Sample that is taken during the use of tap water for human consumption in a property. The sample provides an approximation of the average concentration of metals that are consumed in that property.
Random daytime sample	Sample that is taken from a randomly selected property at a random time of the working day, directly from the tap in the property without previous flushing.
Stagnation sample	Sample that is taken after prolonged flushing of the tap in a property (see fully flushed sampling) and subsequent stagnation for a predefined period before the sample is taken from the tap. During the stagnation period no water should be drawn from any outlet within the property (this includes flushing of toilets).
Water Safety Plan	A Water Safety Plan is a comprehensive risk assessment and risk management approach that encompasses all steps in water supply from catchment to consumer. It comprises as a minimum the three essential

actions that are the responsibility of the drinking-water supplier in order to ensure that drinking-water is safe. These are: a system assessment, effective operational monitoring, and management.

Limit of detection — The lowest quantity of a substance that can be distinguished from the absence of that substance within a stated confidence limit. It is calculated as three times the relative within batch standard deviation of a natural sample containing a low concentration of the parameter, or five times the relative within batch standard deviation of a blank sample.

Precision — The random error and is usually expressed as the standard deviation (within and between batch) of the spread of results about the mean. Acceptable precision is twice the relative standard deviation.

Trueness — The systematic error and is the difference between the mean value of the large number of repeated measurements and the true value.

Foreword

Iron and manganese are both extremely common elements in the natural environment and it is not surprising that they are often present in significant concentrations in water supplies. Since Roman times, miraculous health-giving properties have been attributed to mineral-rich springs containing high concentrations of iron, manganese and other elements. Throughout Europe and America in the 18th and 19th centuries, whole resorts sprang up around these chalybeate springs as people flocked to "take the waters" for their many supposed benefits, including curing colic, flatworms and numerous other ailments, as well as acting as a general tonic.

In the 21st century, consumers are less willing to tolerate high concentrations of iron and manganese in their drinking water due to the effect these substances have on the appearance and taste of the water. In Scotland, in 2010, over 14,000 people complained about discoloured water supplies, making up 70% of all quality complaints received by the water supplier. This is probably a typical figure for a water supplier with a large quantity of ferrous water mains, but some water suppliers have demonstrated that regulatory non-compliance and consumer dissatisfaction can be significantly reduced via targeted investment and enhanced operational strategies.

Regulatory standards for iron and manganese around the world are usually set from the point of view of avoiding adverse aesthetic impacts. Although neither substance is generally considered to be of health significance at the concentrations found in water supplies, this should not be taken for granted and there has recently been renewed interest in the potential effects of manganese on IQ development within children, which will no doubt attract further study in years to come.

Clearly, it is vital that water supply practitioners are actively managing the issues of iron and manganese within their water systems and this Best Practice Guide provides a comprehensive and practical reference to the subject, covering the occurrence, monitoring and control of these substances.

Matthew Bower
Operations Team Leader
Drinking Water Quality Regulator for Scotland

Executive Summary

Iron and manganese are among the most abundant elements in the earth's crust. Both metals may be present in the environment in various oxidation states. In water, iron exists mainly as Fe(II) or Fe(III) and manganese as Mn(II) and Mn(IV). Iron and manganese are commonly found in groundwater due to the weathering and leaching of metal bearing minerals and rocks into the aquifers; in confined aquifers that are devoid of oxygen, iron and manganese concentrations can vary by several orders of magnitude and often exceed the parametric values set for drinking waters. Iron and manganese concentrations can also be elevated as a consequence of complexion with natural organic matter, particularly with the humic acids commonly associated with upland surface waters.

Iron is vital for all living organisms, because it is essential for numerous metabolic processes including oxygen transport, DNA synthesis, and electron transport. In humans, iron is accumulated in the liver and spleen. Haemoglobin and myoglobin contain about 70% of the total iron content in the human body.

Manganese is generally regarded as a nuisance rather than a toxic component of drinking water but presents a special problem for risk assessment because it is both an essential nutrient and a potential neurotoxin. Its neurotoxic properties have emerged almost exclusively from inhalation exposures, although some epidemiological data suggest that high concentrations in drinking water may be associated with neurological impairment. Compared to adults, exposure to manganese will be higher in infants and children, for any particular concentration in drinking water, due to the greater water consumption relative to their body weight.

Presently, iron and manganese are not considered as toxic elements by the World Health Organization and no health based guideline values were set in their latest (2011) 'Guidelines for Drinking-Water Quality'.

In many countries, standards for iron are set for aesthetic acceptability reasons, whereas standards for manganese are set for both health and acceptability considerations. The manganese standards set for aesthetic aspects are more stringent than those that would be needed for health protection.

The presence of manganese in raw water presents special problems during water treatment because, unlike iron, manganese is not easily oxidized by air at neutral pH and may not be removed by water treatment processes unless a chemical oxidation step is included. The presence of even low levels of manganese in the water supplied to a distribution system may result in the deposition of manganese oxides on pipes. This may cause subsequent sloughing of manganese oxide deposits and result in poor aesthetic water quality. Such water has a brown-black colour, an undesirable taste and can cause staining of fixtures, equipment, swimming pools, and laundry.

A large proportion of the customer complaints that drinking water supply companies receive stem from the occurrence of iron discoloured water. Such discoloured water may stain household appliances, porcelain ware and laundry. Discoloured water incidents may adversely affect the customers' confidence in tap water quality.

Corrosion in water distribution systems can be a major problem in water supplies. Corrosion may destroy the fabric of pipes, use up oxidants and disinfectants in water, and the formation of deposits can cause hydraulic restrictions that increase the amount of energy needed for the pumping of water to consumers. Corrosion will assist the development and maintenance of biofilms and may result in the formation of suspensions composed of iron particles that cause red, brown or yellow colouration of the water.

In many countries, cast iron and other iron-based materials are the most commonly used materials for water mains and service connection pipes. The cost of replacing or lining such pipes can cost many billions of dollars (US) or euros (EU).

A holistic strategy for controlling iron and manganese in a water supply system must embrace regulatory compliance, levels of service to consumers and asset management. Developing a strategy will require risk assessment that is integrated with routine performance measurements. Any deficiencies with source waters should be resolved before tackling any problems within the distribution network. Engineering measures must be properly determined by project appraisal before commencement. Monitoring must be sufficient for determining the nature and extent of any problems within the system, and for providing a robust basis for corrective actions. The strategy should be reviewed annually and amended as necessary.

Chapter 1

Introduction: Why there are problems with iron and manganese in water supply

1.1 NATURE OF IRON AND MANGANESE

Iron and manganese are among the 10 most abundant elements in the earth's crust. Iron is the most common element (by mass) forming the planet Earth as a whole.

Both metals may be present in environment in various oxidation states. In water iron exists mainly as Fe (II) or Fe(III) and manganese as Mn(II) and Mn(IV). In anoxic conditions, such as in ground waters or the hypolimnion of eutrophic reservoirs, iron and manganese occur in relatively soluble +2 oxidation state. When exposed to oxygen or disinfectant during water treatment and distribution, Fe (II) is oxidized to the insoluble Fe(III) form, which precipitates and causes the release of iron from distribution system.

The presence of manganese in raw water presents special problems during water treatment because, unlike iron, manganese is not oxidized by air at neutral pH and is not removed during water treatment processes unless a chemical oxidation step is included. The presence of even low levels of manganese in water supplied to distribution system may result in deposition of manganese oxides on pipes. This may cause sloughing of manganese oxide deposits and result in poor aesthetic quality of water. The water has a brown-black colour and undesirable taste and which causes staining of fixtures, equipment, swimming pools, and laundry.

Iron is vital for all living organisms because it is essential for multiple metabolic processes to include oxygen transport, DNA synthesis, and electron transport. Iron is accumulated in the liver and spleen. Haemoglobin and myoglobin contain about 70% of total iron content in a human organism.

Like iron, manganese is regarded as a nuisance rather than a toxic component of drinking water but presents a special problem for risk assessment because it is both an essential nutrient and a potent neurotoxicant. Its neurotoxic properties have emerged almost exclusively from inhalation exposures, although some epidemiological data suggest that high concentrations in drinking water may be associated with neurological impairment. Reported concentrations ranging from 160 to 1200 μg Mn/l are much higher than aesthetic limits – 50 μg Mn/l, suggested by WHO and implemented in drinking water directive.

Iron and manganese are commonly found in groundwater due to the weathering and leaching of manganese bearing minerals and rocks into the aquifers; concentrations can vary by several orders of magnitude and very often exceed parametric values set for drinking waters. Figure 1.1 present an example from Poland. Only 30% of samples present concentrations of iron below 0.2 mg/l and 25% of samples show concentrations of manganese below than 0.05 mg/l. This means that about ¾ of groundwater require iron and manganese removal before it will meet aesthetic limits and can be used for human consumption.

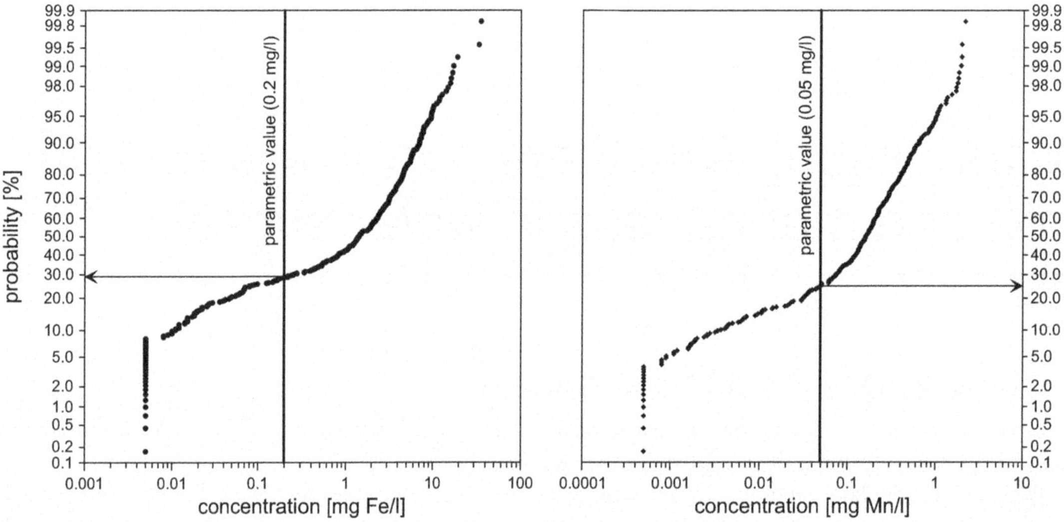

Figure 1.1 Concentrations of iron and manganese in groundwater in Poland. On the base of 2011 groundwater quality monitoring results.

In surface waters iron is generally present as salts containing Fe(III) when the pH is above 4.5. Most of those salts are insoluble and settle out or are adsorbed onto surfaces; therefore, the concentrations of iron in well aerated waters are seldom high.

Manganese is generally present in natural surface waters as dissolved or suspended matter at concentrations below 0.05 mg/l.

However, in some surface waters, particularly in upland catchments and reservoirs increased concentrations of iron and manganese may present serious problems. The UK uplands provide headwaters of many major British rivers and they are a major source of potable, industrial and agricultural water supplies. The main focus on upland water quality has been on those components directly linked to acidification and forestry rotation cycles (pH, aluminium, chloride, sulphate and nitrate). Recent studies from UK have shown that concentrations of iron and manganese can be high in upland waters (Abesser *et al.* 2006).

An additional problem associated with lakes and ponds is stratification and seasonal turnover. During summer stratification, temperature and dissolved oxygen decrease with increasing water depth and correlate with the production of reduced soluble chemical species of iron, manganese and ammonia and hydrogen sulphide in the hypolimnion or anoxic bottom waters of the lake, as a result of the decomposition of organic matter. In late autumn when the temperature of the epilimnion decreases the density difference and wind forces acting on the surface are powerful enough to bring about a reversal of the layers allowing the water from hypolimnion to rise to the surface. Seasonal temperature variations additionally modify the vertical structure and red-ox conditions. Iron, manganese and other compounds (i.e. sulphur as hydrogen sulphide) may become increasingly mobile and when released from anoxic bottom sediments may cause taste and odour problems. Fluctuations in concentrations of iron and manganese may affect treatment process and as a result of insufficient treatment lead to customers complains.

1.2 CONSUMERS' PERCEPTION

A large proportion of the customer contacts that drinking water supply companies receive stem from the occurrence of discoloured water. "Coloured water" describes the appearance of drinking water that contains suspended particulate iron or manganese where the actual suspension colour may range from light yellow to red to brown and brown-black due to water chemistry and particle properties. The coloured water stains household appliances and porcelain ware. Clothes laundered in such water are also stained.

Discoloured water incidents greatly affect customer's confidence in tap water quality and the quality of service provided by water companies. Currently, consumer complaints are dealt with in a reactive manner. However, water companies are being driven to implement planned activities to control discolouration prior to contacts occurring. Water discolouration was, in early 1990's, the single largest cause of complains in the Yorkshire water region (Pattison *et al.* 1995). The proportions of complaints have changed significantly in many English companies since 1995, with discolouration no longer the largest component, although this has taken extensive investment in the rehabilitation of iron water mains over a number of years. According to Vreeburg & Boxall (2007) water discolouration is the reason for 34% of customer contacts with Water Company in UK.

Water with unacceptable colour will undermine the confidence of consumers which could lead to the use of water from sources that are aesthetically more acceptable, but potentially less safe. The consumers however, have no means of judging the safety of their drinking water themselves. Their attitude towards their drinking water supply and their drinking water suppliers will depend on aspects that they will be able to perceive with their own senses.

In general, number of complaints increase with increasing distance from the treatment plant. This is in part due to the fact that the velocity of flow decreases with increase in distance from the plant. The iron dissolved or the decrease in dissolved oxygen per unit distance travelled by a unit volume of water near the plant is far less than that at the outer ends of the distribution system. The time of contact with the pipe per unit volume of water is relatively short for the larger mains and longer for the smaller pipes where the demand is low or negligible for significant periods of time. Furthermore, as corrosion takes place and rust particles develop, they settle to the bottom of the pipe at periods of low or negligible demand, and are picked up and redistributed at periods of high or instant demand.

1.3 ECONOMIC AND TECHNICAL PROBLEMS

Treatment

Traditional technology of water treatment is based on transformation of Fe(II) and Mn(II) compounds to the form of insoluble oxides of Fe(III) and Mn(IV) that are easy to be removed by filtration.

Manganese removal requires oxidation of Mn (II) by two degrees of valence and higher redox potentials which makes this process technologically more demanding than iron removal. In order to intensify the process of manganese oxidation stronger oxidation conditions have to be used than oxygen at neutral pH. This may be achieved by the use of stronger oxidising agents such as chlorine, chlorine dioxide, potassium permanganate and ozone, or the use of autocatalytic or biological processes.

Corrosion in distribution network

Corrosion in water distribution systems is one of the most important problems in water supply. In many countries worldwide cast iron and other iron-based materials are the most commonly used materials for mains and service pipes.

In the United States approximately 15% of all pipelines supplying drinking water are unlined cast iron pipes (AWWA 2003). The American Water Works Association estimates that in order to upgrade water distribution systems in United States, \$ 325 billion should be spent over the 20 years period (AWWARF 1996). This AWWA value is built on the United States Environmental Protection Agency (US EPA) estimate of U\$ 77.2 billion for service and replacement of transmission and distribution system lines over the 20 years period (Davies *et al.* 1997). Latest estimations are even higher reaching \$526 billion during period 2011–2035 (AWWA 2011).

The majority of distribution system pipes are composed of iron material: cast iron (41%), ductile iron (25%), and steel (3%) (AWWA 2011). According to American Water Works Association Research Foundation corrosion of cast iron pipes is the most common distribution system problem (AWWARF 1996). Corrosion of iron pipes in a distribution system can cause many problems such as:

(1) Pipe mass is lost through oxidization to soluble iron species or iron-bearing scale.
(2) The scale can accumulate as large tubercles that increase head loss and decrease water capacity (see Figure 1.2a).
(3) Localised corrosion leading to pipes breaks (see Figure 1.2b).
(4) The build-up of corrosion products into tubercles inside the pipes may provide sites for microbial regrowth.
(5) Corrosion reactions can also directly consume disinfectants, thereby lowering residuals and allowing increased biological activity. Harmful disinfection by-products such as chloroform may be formed as chlorine disinfectants react with tubercles in corroded iron pipes.
(6) The release of soluble or particulate iron and manganese corrosion by-products to the water decreases its aesthetic quality and often leads to consumer complaints of "red water" or "black water" at the tap.

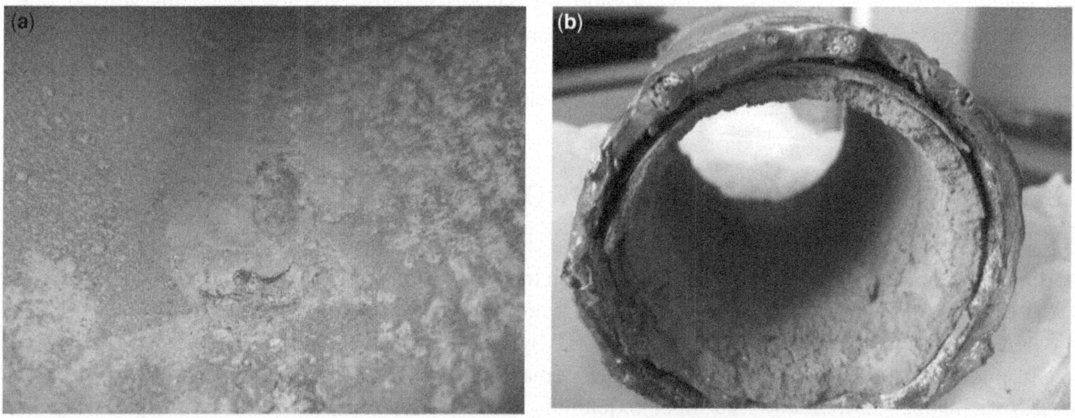

Figure 1.2 Effects of iron corrosion in distribution system; (a) localised pitting corrosion, (b) corrosion scales inside iron pipe.

Corrosion of iron pipe is a complicated process that is influenced by many different factors. The most important are:

- Water quality and composition (DO, pH, redox potential, chlorides, sulphates).
- Flow conditions.
- Biological activity.
- Pipe age.
- Temperature.
- Corrosion inhibitors usage.

Concentrations of chemical constituents affecting the rate of corrosion can change during passage through the system. A certain water may be only slightly corrosive in one part of a system and strongly corrosive in another due to the difference in time of contact for corrosion to take place per unit volume of water.

Corrosion scales and deposits formed within drinking water distribution systems have the potential to retain inorganic contaminants.

Dissolved oxygen in water system gives rise to conditions which may prevent or enhance "red water". Inhibition is experienced by the formation of a protective ferric oxide calcium carbonate coating in the mains which can prevent penetration of dissolved oxygen to the metal itself as well as stop corrosion products from sloughing off.

The presence of ammonia and oxygen as a source of energy for bacteria gives rise to bacterial growths in the mains. Corrosion is enhanced by the bacterial transformation of basic ammonia and oxygen to acidic and oxidizing nitrites and nitrates.

Depletion of dissolved oxygen at the "dead" ends and service lines of the system gives rise to a condition where ferrous iron is not oxidized to insoluble ferric oxide and solution of iron to the ferrous state takes place.

KEY POINTS:

(1) Iron and manganese are commonly found in groundwaters and some surface waters due to the weathering and leaching of iron and manganese bearing minerals and rocks under anoxic conditions; concentrations can vary by several orders of magnitude.

(2) Both iron and manganese may be present in environment in various oxidation states. In water iron exists mainly as Fe(II) or Fe(III) and manganese as Mn(II) and Mn(IV).

(3) Effective manganese removal is more difficult than iron removal and requires usage of stronger oxidizing conditions.

(4) The presence of even low levels of manganese in water supplied to distribution system may result in deposition of manganese oxides on pipes deposits and result in poor aesthetic quality of water.

(5) Corrosion of iron pipes and release of soluble or particulate iron corrosion by-products to the water decreases its aesthetic quality and often leads to consumer complaints.

Chapter 2

Natural sources of iron and manganese in water

2.1 HYDROGEOCHEMISTRY OF IRON AND MANGANESE

Iron

It commonly occurs in the Earth's crust (4.65%) and the primary sources of iron in groundwater and surface water are the magmatic rock minerals such as: pyroxenes, amphiboles, magnetite, olivines and pyrite. In these minerals iron mainly occurs in the bivalent form. As a result of the process of magma rock decay, iron gets to the environments of sediment rocks, where it occurs mainly in the form of oxides and hydroxides (hematite, limonite, goethite), as well as pyrites and siderites.

The migration of iron into natural water occurs as a result of decay with the participation of water and CO_2, as well as oxidation and hydrolysis (Macioszczyk 1987).

In groundwater, iron usually occurs in the bivalent form, while in surface water it is in the form of a colloidal suspension of trivalent iron. The content of iron in groundwater is mainly determined by its form in sediments – Fe(II) or/and Fe(III), the pH of the environment and the reduction-oxidation potential (Eh). An example of such relationships, based on observations from western Poland, is presented in Figure 2.1 after Górski (1981).

The trivalent iron is practically insoluble in water at the usual pH encountered. According to the nature of iron release into water, the environments characterized by the following conditions can be distinguished: oxidizing, transitional and reductive.

In the oxidizing environments with Eh > 400 mV (at pH > 7), iron transforms into the oxidized form Fe(III) and it does not occur in groundwater. It also does not occur in the water in highly reductive environments rich in organic matter. In these environments, as a result of anaerobic decomposition of organic matter, bivalent sulphur ions are released, which, when combined with iron, are precipitated in the form of insoluble sulphides. High concentrations of iron occur in the environments characterized by a relatively low redox potential (due to the content of organic matter) and acidification connected with the presence of carbonic acid and humic acids.

Moreover high concentrations of iron in water are observed in the oxidation zones of sulphide deposits, where sulphuric acid is formed as a result of the oxidation of sulphides.

Groundwaters usually contain the bivalent iron, which undergoes oxidation and turns into colloid suspension. Iron in the colloid form and the suspension form also get into rivers and reservoirs as a result of surface run-off.

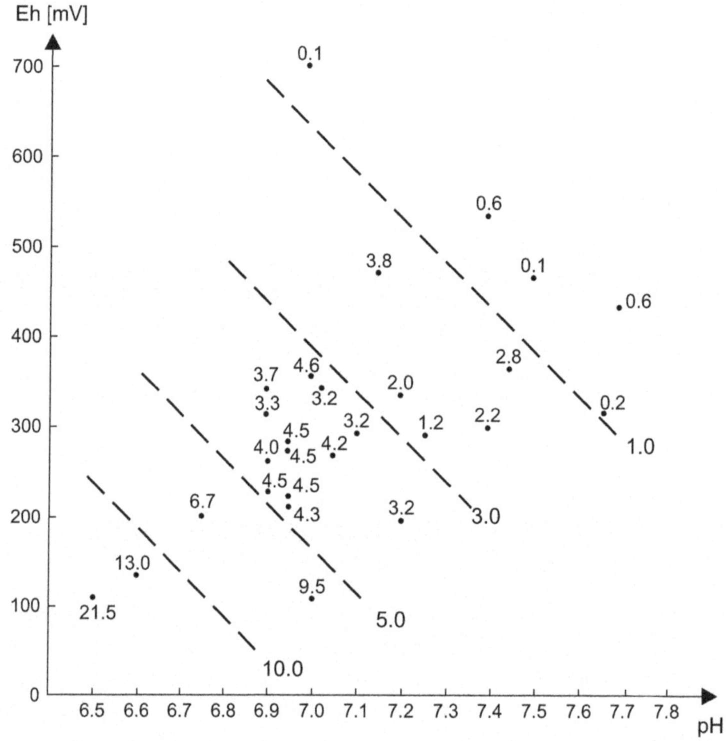

Figure 2.1 Iron concentrations (in mg/l) in groundwater depending on the pH and Eh. Quaternary aquifers in western Poland (Górski 1981).

The main source of iron in surface water is connected with groundwater drainage to rivers and lakes. It may also be generated by erosion and the dissolution of iron and/or manganese containing sediments and minerals at the sediment-water interface.

It was reported that in some rivers and stream waters in Sweden iron was present in concentrations up to 1.7 mg/l accompanied with manganese in concentrations up to 0.14 mg/l (Huser *et al.* 2011; Kritzberg & Ekström 2012). In the UK, uplands provide headwaters of main British rivers which are a major source of drinking water. Concentrations of iron occurring in stream and river waters have doubled over the past 20 years. In some small streams water contains up to 4.7 mg Fe/l and in main streams up to 0.8 mg Fe/l (Neal *et al.* 2008).

Manganese

Manganese occurs in the Earth's crust in amounts much smaller than iron (0.1%). In magmatic rocks, it occurs in the dispersed form as an admixture of dark minerals in olivines, pyroxenes, amphiboles and biotites. It occurs as Mn(II), Mn(III), Mn(IV). The trivalent form is transient and unstable. From magma rocks, it is mainly released as Mn(II) and it is more easily leached than iron. Oxidation to Mn(IV), which is very weakly soluble and is produced in water as MnO_2, requires stronger oxidizing conditions than the oxidation from Fe(II) to Fe(III).

In surface water, manganese migrates mainly in the form of colloid suspension, while in groundwater it migrates as Mn(II), and its concentrations are usually about ten times lower than those of iron.

The main factors determining the solubility of manganese minerals and its migration in groundwater are Eh and pH. The migration is also facilitated by forming complexes with organic matter. The highest concentrations of manganese of up to 5–7 mg/l are observed in alluvial formations rich in organic matter on the flood terraces of valleys and ice-valleys. High manganese concentrations are also found in the zones of sulphide deposits oxidation, similarly as those of iron.

2.2 VARIABILITY OF CONCENTRATIONS OF IRON AND MANGANESE IN GROUNDWATER

The occurrence of iron and manganese is a typical feature of groundwater and it concerns the majority of wells. According to the data collected in the regions of Northern and Central Poland, 71.5% of wells reach the Fe contents of above 0.2 mg/l. That is above the allowable level according to the Polish sanitary regulations and 76.8% of the well-fields reach the manganese concentrations of above 0.05 mg/l (Table 2.1).

Table 2.1 Occurrence of iron and manganese (in mg/l) in groundwater in the regions of Northern and Central Poland.

	Aquifers in deposits															Total
	Quaternary					Paleogene and Neogene					Cretaceous and Jurassic					
	1	2	3	4	5	1	2	3	4	5	1	2	3	4	5	5
Iron	3346	1.85	0.99	50.4	70	716	1.94	1.01	29.7	80	767	1.25	0.7	18.1	70	71.5
Manganese	3336	0.21	0.14	5.81	80	716	0.14	0.11	1.7	80	767	0.09	0.06	3.6	60	76.8

1 – no. of analyses; 2 – arithmetic mean; 3 – median; 4 – maximum value; 5 – percentage of analyses above the parametric value (0.2 mg Fe/l and 0.05 mg Mn/l).

The highest concentrations of iron and manganese within groundwater are observed in Quaternary, Palaeogene and Neogene units which mainly consist of confined aquifers. Significantly lower concentrations of the ions are found in Cretaceous and Jurassic fissure aquifers under partly a confined or semi confined conditions.

In unconfined aquifers, especially when water is highly contaminated with nitrates, lack of iron and sometimes also manganese is often observed. Similar low content or even absence of the Fe and Mn is observed in aquifers with thick aeration zones (see Figure 2.2).

Such a situation is the effect of long-term influence of atmospheric oxygen, which leads to the full oxidation of iron and manganese. However, such a situation more frequently relates to iron, as the full oxidation of manganese requires higher oxidation potential. That is the reason why at the shallow aquifers the lack of iron is observed with the occurrence of manganese. The zones of full oxidation of iron and manganese compounds can be formed when these formations do not contain large amounts of organic matter.

In the case when water-bearing formations are enriched with dispersed organic matter, which especially concerns the recent alluvial formations on the flood terraces of valleys and ice-marginal valleys, favourable conditions to release iron and manganese compounds to water and their migration are created. This is related to the influence of organic matter on the lowering of Eh. Moreover, there are conditions favourable for the lowering of pH as a result of the influence of humic acids, as well as the formation of carbonic acid as a result of organic matter oxidation.

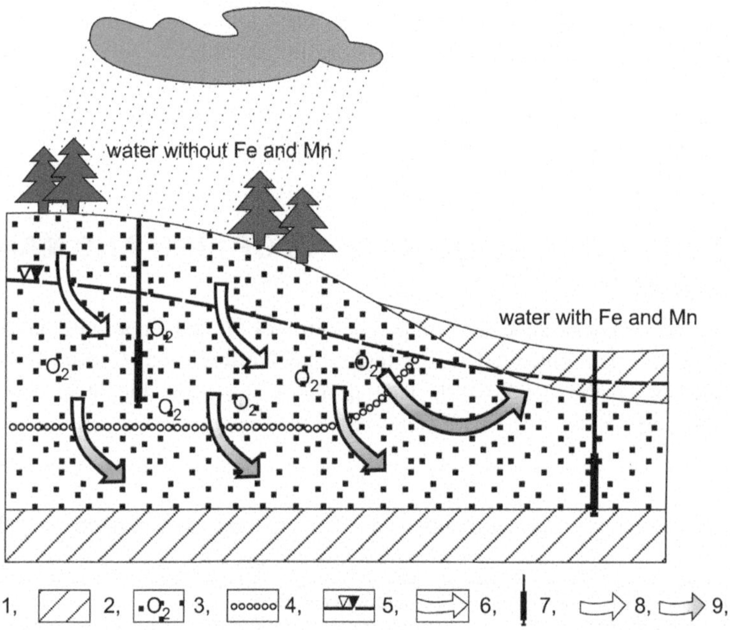

Figure 2.2 Conditions of the occurrence of water not containing iron and manganese at the shallow aquifer (Górski 2011). 1 – water-bearing formations-sands, 2 – non-water-bearing formations – clays, 3 – fragment of water-bearing layer where oxidation conditions occur, 4 – oxygen zone range, 5 – water table, 6 – water flow directions, 7 – well screen, 8 – water not containing iron and manganese, 9 – water containing iron and manganese.

The above situation has been illustrated by the data presented in Figure 2.3, where the highest Fe concentrations are clearly related to the regions of facilitated infiltration of precipitation water, and where the favourable conditions for the oxidation of organic matter gathered in the alluvial formations on the flood terrace occur.

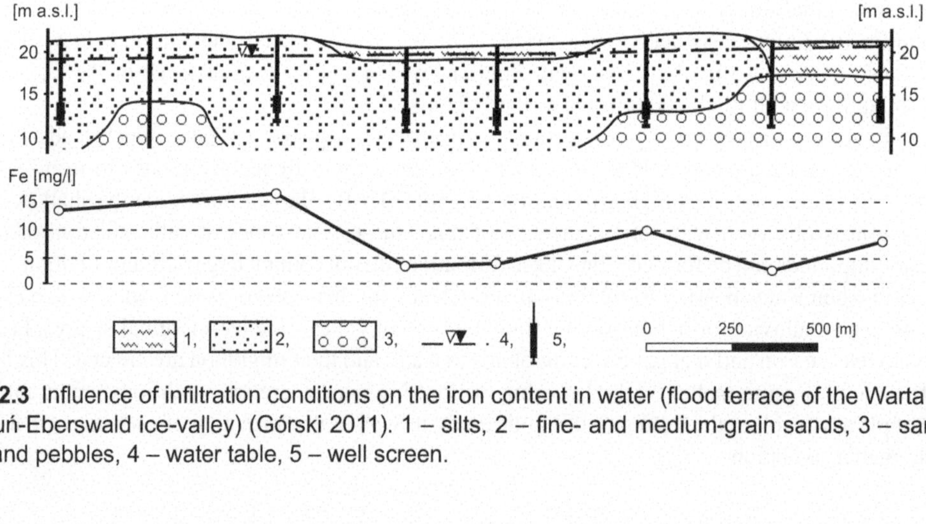

Figure 2.3 Influence of infiltration conditions on the iron content in water (flood terrace of the Warta River in the Toruń-Eberswald ice-valley) (Górski 2011). 1 – silts, 2 – fine- and medium-grain sands, 3 – sands with gravel and pebbles, 4 – water table, 5 – well screen.

It should be added that, to some extent, the high concentrations of Fe can also be related to the oxidation of iron sulphides, which leads to the high acidification with sulphuric acid. The presence of iron sulphides is linked to the anaerobic decomposition of organic matter and the release of S^{2-} ions which, when combined with Fe(II), form the colloid FeS. The influence of this process is, however, most noticeable in the case of the hydrodynamic changes connected with water pumping, which is presented in the next chapter.

The occurrence of high concentrations of iron and manganese in formations enriched with organic matter is also related to the conditions of water exchange. Particularly high concentrations can be observed in the environments where the conditions of water stagnation occur. An example of such an environment is presented in Figure 2.4, where at the region between two river beds of the Oder, in the conditions of water stagnation, the Fe concentrations of up to 80 mg/l and Mn concentrations of up to 2 mg/l were found.

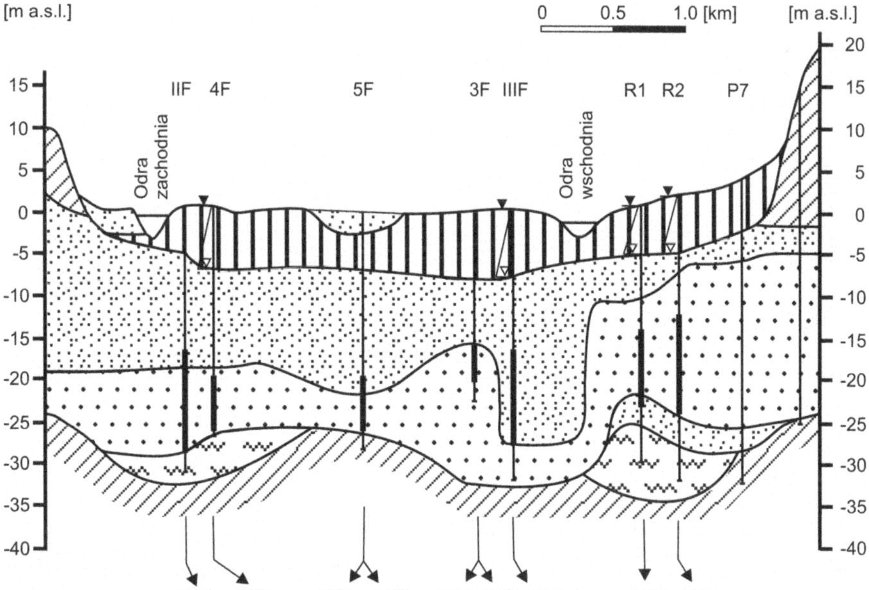

Date	1984	1987	1984	1987	1984	1987	1984	1983	1989
Color [mg Pt/l]	60	35	55	35	165	35	140	10	15
Oxidation [mgO₂/l]	6.5	17.8	49	29.6	95	29.6	26	4.8	2.2
Cl^- [mg/l]	25	19.8	39.5	60.4	25	120	19.5	21.5	24
SO_4^{2-} [mg/l]	44.5	16.9	16.9	33.3	2.0	4.9	4.1	16.0	14.8
N_{NH_4} [mg/l]	2.6	4.5	3.0	3.5	4.0	6.0	15.0	0.45	0.6
N_{NO_3} [mg/l]	0.1	BLD	0.2	BLD	0.3	BLD	0.5	BLD	BLD
Fe [mg/l]	8.0	32.0	50	42	60	68	80	4.0	4.0
Mn [mg/l]	0.65	0.9	0.9	1.0	1.5	1.3	2.0	0.1	-

BLD - below limit of detection

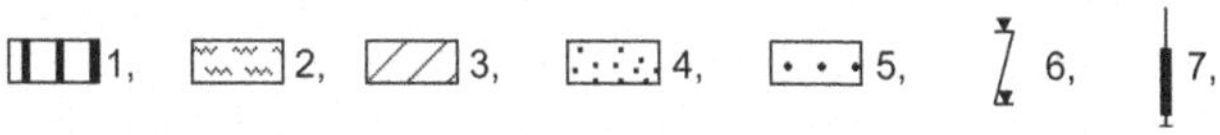

Figure 2.4 Example of extremely unfavourable hydrogeochemical environment in the Lower Oder valley (Górski 2004). 1 – peats, 2 – silts, 3 – glacial tills, 4 – fine- and medium-grained sands, 5 – sands with gravel, 6 – water table, 7 – well screen.

In the aquifers isolated from the direct influence of atmospheric factors, the variability of Fe and Mn concentration is lower, where in the case of Fe the general increase of mean concentrations depending on the depth is observed, which results from the lowering of Eh (see Figure 2.5).

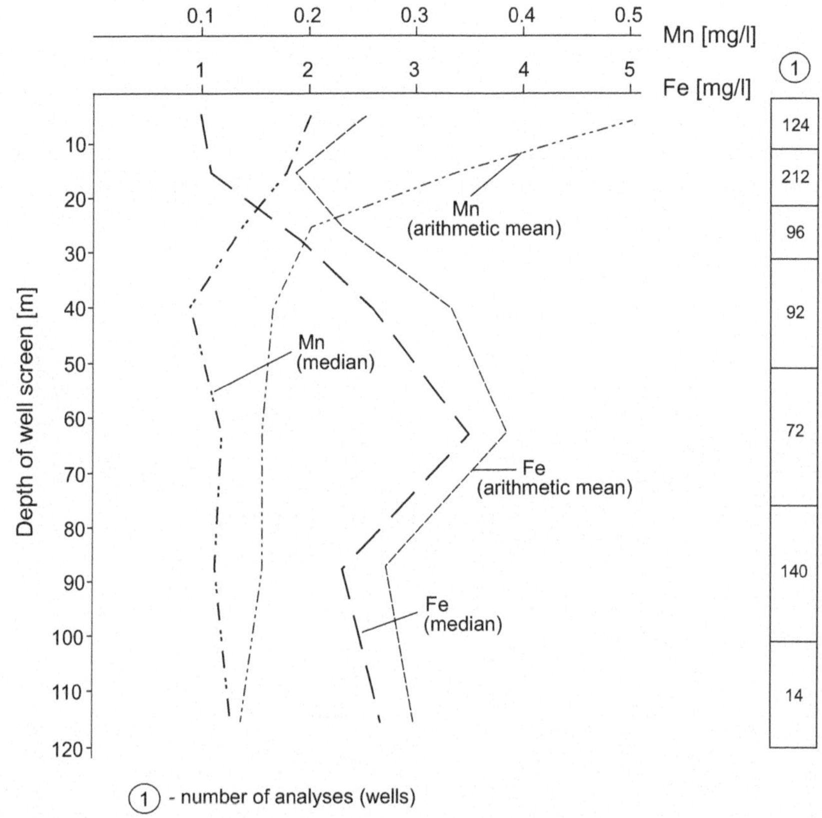

Figure 2.5 Variability of average iron and manganese concentrations in groundwater depending on the depth of well screen location (aquifers in Quaternary deposits in Wielkopolska) (Górski 1989).

On the other hand, the mean concentrations of manganese are higher in shallow aquifers, and in the deep aquifers they do not increase depending on the depth. In the case of iron, we can observe the variability connected to the redox conditions. Here, lower concentrations in the zones formerly subjected to long-term influence of oxygen (buried aeration zones) can be observed (see Figure 2.6).

The oxidizing conditions in such zones are created due to the phenomenon of oxygen adsorption on trivalent iron hydroxides, which leads to the iron removal from water flowing from outside these zones (Górski 2011).

Also the lithology of water-bearing formations to some extent influences the variability of iron concentrations in deep water. Within the area of post-glacial Quaternary formations, lower concentrations are observed in the aquifers formed from coarse-grained fluvio-glacial formations which contain carbonates. Less favourable conditions occur in fine-grained fluvial formations, especially when these are rich in organic matter.

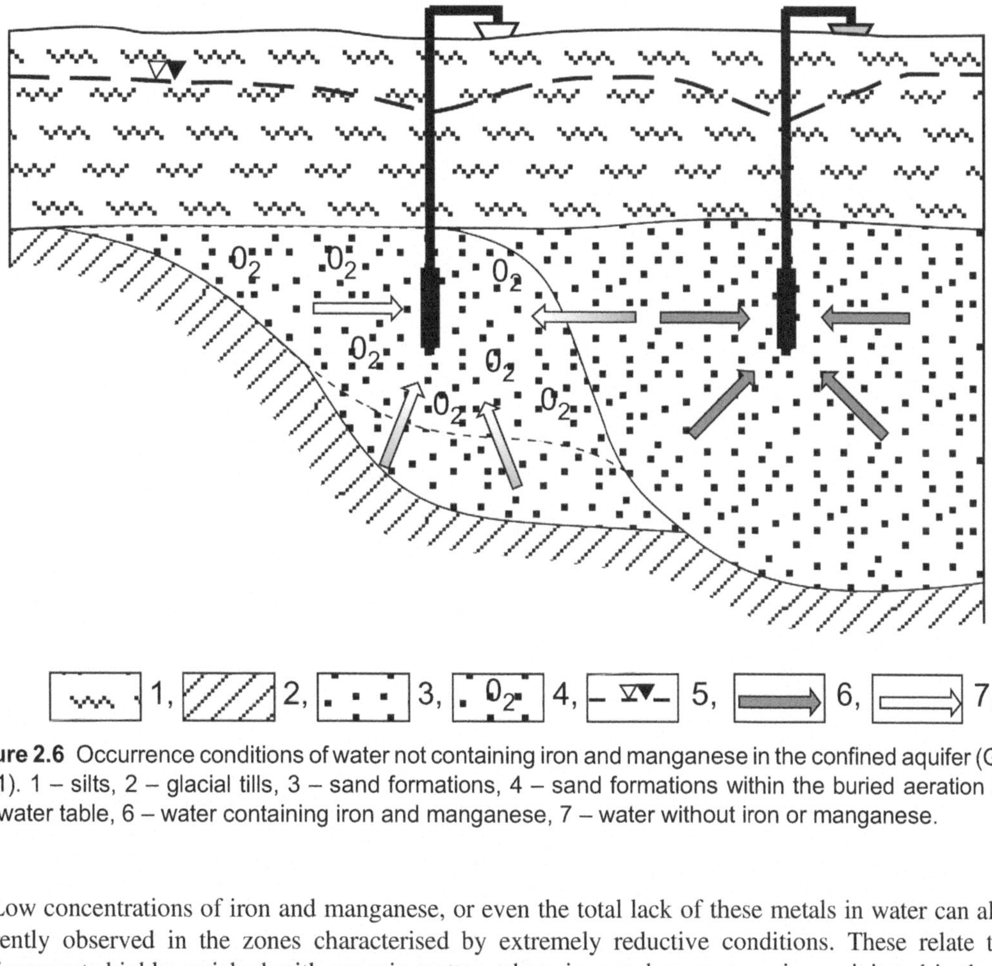

Figure 2.6 Occurrence conditions of water not containing iron and manganese in the confined aquifer (Górski 2011). 1 – silts, 2 – glacial tills, 3 – sand formations, 4 – sand formations within the buried aeration zone, 5 – water table, 6 – water containing iron and manganese, 7 – water without iron or manganese.

Low concentrations of iron and manganese, or even the total lack of these metals in water can also be currently observed in the zones characterised by extremely reductive conditions. These relate to the environments highly enriched with organic matter, where iron and manganese is precipitated in the form of insoluble sulphides. The occurrence of such an environment has been found locally within the brown coal Miocene formation in Wielkopolska (Górski 1989).

The manganese concentrations are highest in shallower aquifers, which are connected with the influence of acidifying factors. In deeper aquifers, the concentrations are more equalized and the influence of the redox conditions variability is less noticeable than in the case of iron.

Lower manganese concentrations in water (<0.1 mg/l) may, on the other hand, occur in certain geochemical environments with a lack of minerals containing manganese. Such situations have been observed in certain zones of the Miocene aquifers in Wielkopolska, where the ascent of water characterised by very low hardness (<4 mg/l) from the formations of Lower Jurassic is found (Górski 1989; Siepak *et al.* 2010).

It should be added that iron and manganese in water are also locally affected by anthropogenic pollutants. Lower concentrations and even the lack of these metals in water are observed in the zones contaminated with nitrates which oxidize the groundwater environment. High concentrations can, on the other hand, occur in the zones enriched with organic matter, for example, in the zones of leachates from landfill sites.

2.3 CHANGES IN IRON AND MANGANESE CONCENTRATIONS DURING GROUNDWATER EXPLOITATION

In the conditions of groundwater pumping, an increase in iron and manganese concentrations is often observed. This phenomenon is especially noticeable when the water is taken from unconfined or poorly confined aquifers in valleys and ice-marginal valleys.

The causes of the concentration increase are as follows:

- Changes in the water circulation system.
- Decrease of the water table level, which may generate the development of unfavourable hydrogeochemical processes.

Changes in the water circulation system result from the fact that in post-glacial river valleys, the wells have screens in their lower parts, consisting of coarse-grained fluvio-glacial formations. The Fe concentrations in this zone usually fall within the range of 2–3 mg/l, and manganese concentrations range from 0.2 to 0.3 mg/l.

Much higher concentrations of iron and manganese may occur in the upper parts of valley deposits, where more fine-grained sediments enriched with organic matter occur. However, in natural conditions, water from the shallower parts does not reach the deeper parts due to the ascending flow of water from the lower to the upper parts. This situation changes in the conditions of exploitation and water from the upper parts reaches the lower ones, which affects the increase in Fe and Mn concentrations. The example illustrating this phenomenon is presented in Figure 2.7 (Górski 1981).

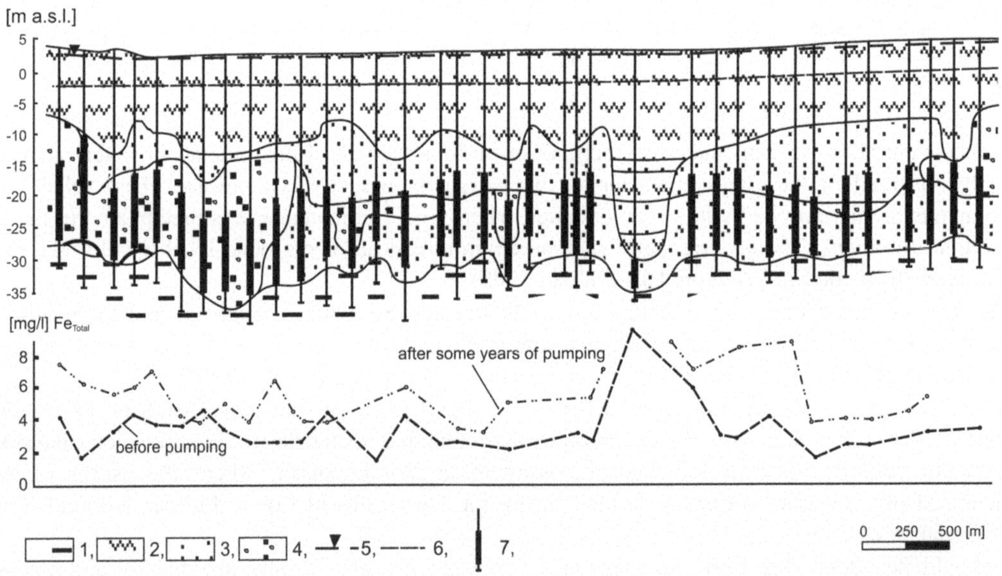

Figure 2.7 Increase of iron concentrations in groundwater in the exploitation conditions resulting from the changes in the water circulation system (the Vistula River Delta region) (Górski 1981). 1 – clays, 2 – silts, 3 – fine- and medium-grained sands, 4 – sands with gravel and pebbles, 5 – water table before exploitation, 6 – water table during exploitation, 7 – well screen.

It should be added that the cause of increases in iron and manganese concentrations can be similar in the case of water pumping from fissured aquifers. Such an increase results from the flow of water (in the exploitation conditions) enriched with iron and manganese from shallow, unusable water-bearing formations to the deeper parts of the aquifers.

In the case of deep, well isolated aquifers, the increase in iron and manganese concentrations is not observed if the aquifers are characterised by homogenous hydrogeochemical environments (e.g. buried sander aquifers).

The increase in iron and manganese concentrations in deep aquifers can be observed in the case of heterogeneity of hydrogeochemical environments resulting from the diversified lithology and the degree of enrichment in organic matter. Such changes are most often observed in the case of taking water from the aquifers in post-glacial buried valleys, where lithological variability of sediments occurs, that is where coarse-grained fluvio-glacial formations and fine-grained fluvial sediments with the content of organic matter can be found. However, the scope of changes here is much smaller than for the unconfined aquifers in valleys and ice-marginal valleys.

A significant factor in lowering the quality of water in some wells can be the phenomenon of changes to the hydrogeochemical environment in the exploitation conditions. This phenomenon especially develops at the wells located in post-glacial valleys and ice-marginal, where at the surface layer, recent, sand alluvial formations with some content of dispersed organic matter occur (Błaszyk & Górski 1981; Górski 1981). Due to the system of water circulation and also because they are often covered by practically impermeable mud, silt and peat, these formations are naturally isolated from the influence of atmospheric conditions.

In such conditions, the decomposition of organic matter leads to the release of bivalent sulphur ions, which, combined with iron dissolved in water, form iron sulphides. Undertaking the groundwater exploitation in such conditions causes the permanent lowering of the water table, which consequently leads to expanding and often also forming the aeration zone. In the formed aeration zone, the processes of sulphides oxidation begin to develop, which leads to the release of bivalent iron and the formation of sulphuric acid, in accordance to the following reaction:

$$2FeS_2 + 7O_2 + 2H_2O \xrightarrow{\text{cat.biol.}} 2Fe^{2+} + 4H^+ + 4SO_4{}^{2-}$$

This reaction is catalysed by sulphur bacteria, and particularly by the *Thiobacillus thiooxidans* species. As a result of this reaction, the concentration of iron, manganese, sulphates, as well as water hardness increase, and the pH decreases.

This process may result in very large, and sometimes even catastrophic, increases in Fe concentrations (as much as above 100 mg/l). This phenomenon can be observed at many well-fields, and its development in time and scale of changes in water quality are differentiated and depend on the local hydrogeochemical conditions.

The decrease in water quality as a result of the above process is transitional, but its influence on the water quality can be even observed for several consecutive years. The degree of quality decrease depends in the first place on the content of sulphides in the zone under the influence of atmospheric oxygen. However, even if the sulphides do not occur, the water quality may still decrease in a smaller scale, which is the effect of the oxidation of organic matter, which leads to the formation of carbonic acid. In order to limit the influence of the above phenomenon on the water quality and define its scale, it is necessary to conduct appropriate hydrochemical and geochemical investigation at the stage of designing the well-field and to determine the limits to its exploitation, and especially to the allowable exploitation drawdown. It should be underlined that excluding the well from exploitation only after the phenomenon has been detected, significantly lengthens its duration.

It should also be added that the process of hydrogeochemical changes described above may develop not only as a result of the oxidation of natural sulphides, but it may also occur for sulphides formed in the environment due to introducing pollutants in the region affected by sewage, leachates from landfill sites and infiltration of surface water polluted with sewage. The examples of changes in water quality resulting from this process are presented in Figure. 2.8 (Górski 1981) and in the case studies (see Appendix 1).

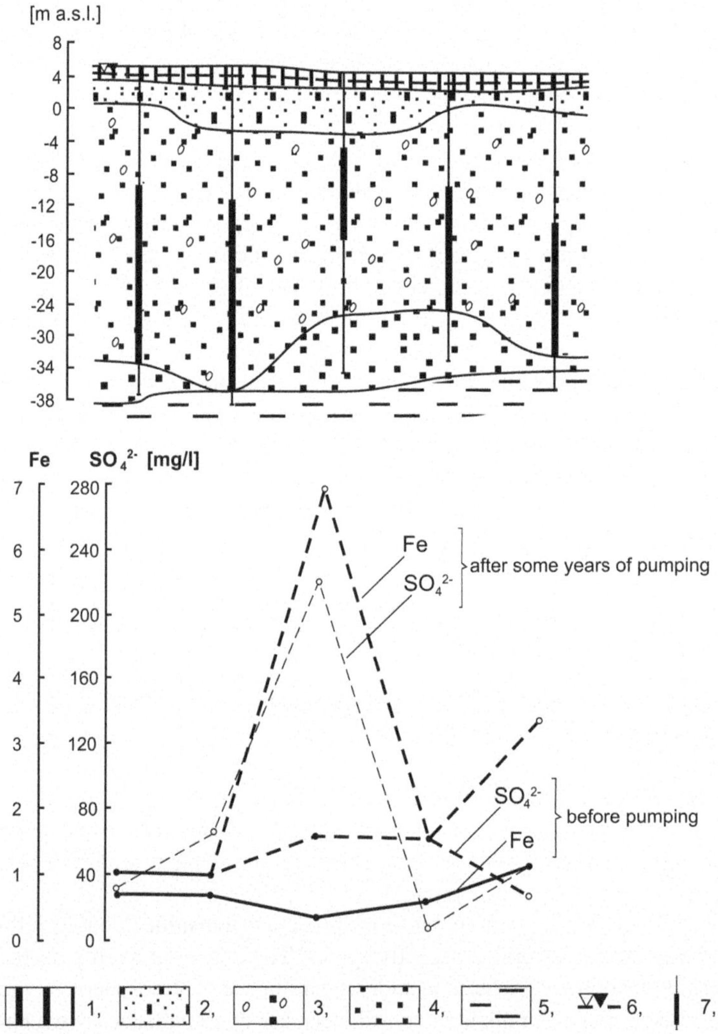

Figure 2.8 Increase in iron concentration in groundwater resulting from the developing processes of sulphides oxidation (the Łeba-Reda ice-marginal valley) (Górski 1981). 1 – peats, 2 – fine- and medium-grained sands containing organic matter, 3 – sands with gravel and pebbles, 4 – fine- and medium-grained sands, 5 – loams, 6 – water table, 7 – well screen.

2.4 SURFACE WATERS

Sources of iron and manganese in surface waters

Iron and manganese occur naturally in many surface waters as a result of catchment erosion and dissolution of minerals containing those metals or near the sediment-water interface. Minerals incorporated into soil from rock weathering are then transformed according to the soil conditions.

Depending on the nature and location of the iron oxidation process, the transport of other chemicals in the same waters may be substantially affected since many substances co-precipitate with iron oxyhydroxides or extensively sorb to iron oxyhydroxide surfaces (Dzombak & Morel 1990; Hamilton-Taylor *et al.* 1996; Hongve 1997). During periods of high flow or storms such iron oxyhydroxide solids and any associated chemicals may be transported downstream (Seeman, 1996; Solo-Gabriele & Perkins, 1997a, b).

In the acidic, organic-rich and moist soils some of the Mn oxides are being reduced to mobile Mn(II) in soil solution which are available for plant uptake and can also be transported by throughflow into runoff. Incorporation of Mn into plant material forms a secondary source of Mn through release from litter decomposition at the ground surface.

Human activities can also be responsible for the iron and manganese contamination of surface water in some areas (Heal 2001). Evidence from Scotland and other parts of the UK indicates that afforestation of upland catchments with conifers since the 1930s has significantly increased Mn concentrations in surface waters. The amount of Mn release depends on the harvest method (Fahey *et al.* 1991a). Mn leaching is higher in plots which are whole-tree harvested than plots which are conventionally harvested. The difference is attributed to the suppression of vegetation regrowth in the conventionally harvested plots with large quantities of logging residue and thus lower uptake of available Mn from the soil. Other mechanisms may also cause increased Mn concentrations after harvesting. These include: Mn release from the decomposition of logging residue, loss of particulate Mn in sediment eroded from harvested areas, mineralisation of Mn in warmer disturbed soils following the removal of the forest canopy and reduced Mn uptake by trees (Fahey *et al.* 1991b). The extent to which the Mn made available by harvesting enters runoff is dependent on catchment hydrology.

The median iron concentration in rivers has been reported to be 700 μg/l (WHO 2003). Recent studies show that iron concentrations in some UK rivers have doubled over the past 20 years increasing for the lower Hafren from average of 71 μg/l to 139 μg/l, reaching maximum of 806 μg/l. In small streams measured concentrations were even higher reaching 4690 μg/l (Neal *et al.* 2008).

Levels of manganese in fresh water typically range from 1 to 200 μg/l (Barceloux 1999). A survey of Canadian surface waters undertaken in 1980–1981 showed the range of manganese to be 10–400 μg/l in freely flowing river water. The highest recorded concentration was 1700 μg/l in the Carrot River in Saskatchewan (Environment Canada 1984). On the base of a river water survey in the USA it was reported that dissolved manganese levels ranged from <11 to >51 μg/l (ATSDR 2000). Some studies from Scotland uplands present much higher values. Deterioration in the Mn quality of raw waters since the 1980s has been reported by a number of water treatment works in the UK which receive runoff from upland catchments (Little & McFadzean 1991). Hardie (*et al.* 2007) reports that during the 1990s, there were three extended periods when water supplied from Loch Bradan to the treatment works contained Mn with peak concentrations of 358–1040 μg/l. A seasonal deterioration of Mn quality in raw waters from the Elan Valley, which supplies water to Birmingham, England, was also reported by Schofield (*et al.* 1991).

Controls on Fe and Mn mobilization

Both iron and manganese can be present in water as a reduced, soluble forms Fe(II) and Mn(II). In contact with atmospheric oxygen, which is typical for aerated surface water, iron oxidize to Fe(III) and manganese to Mn(IV) insoluble forms and usually precipitate as hydroxides. The mechanisms may include homogeneous solution reactions, reactions on surfaces and microbially mediated processes (Emerson & Revsbech, 1994a, b; Ishigaki *et al.* 2005).

Combined effects of pH and redox potential affect most chemical substances, including iron and manganese. When oxygen is present in water it will dominate the redox potential and thus the oxidation state and solubility of all redox-type elements present. Based on theoretical equilibria dissolved iron should not be present in any oxygenated water, but could be very abundant in anoxic, slightly acidic water. In actual practice iron is frequently analytically detectable in surface waters. This is caused by the phenomenon of chelation. Certain complex ions can bind to metallic ions, resulting in a charged complex which remains in the water. In natural waters among the most important chelating agents are organic colour and complex colloid of humic material. Manganese is less readily complexed by organic ligands (LaZerte & Burling 1990). The most common mobilisation mechanism for Mn is probably an abiotic redox reaction, with constituents of the organic matter serving as an electron donor (Kuiters & Mulder, 1993).

Climate changes can also influence the concentrations of iron and manganese in surface waters. Increased mobility of iron was noted in a study of the effects of mild winters and hot summers on stream water chemistry of an upland catchment in mid-Wales (Cannell & Pitcairn 1993). In temperate environments stormier conditions and warmer temperatures may cause increased Mn concentrations in autumn storm events due to flushing of larger loads of mobile Mn, created by enhanced decomposition of organic material in soils in warmer summer temperatures.

Rivers and streams

Due to in-stream processes iron and manganese concentrations can alter after entry into surface waters. The most significant process in flowing water is probably oxidation and precipitation of Fe(III) and Mn(IV) oxides on channel substrate when runoff enriched in those metals encounters well-oxidised conditions in the stream channel. The precipitation of oxides lowers dissolved Fe and Mn concentrations in streamwater and also adsorbs dissolved heavy metal ions (e.g. Cd, Co, Cu, Ni, Pb, Zn).

Although the oxyhydroxides are dominant in well-oxygenated waters, measureable concentrations of Fe(II) and Mn(II) can be observed, especially where photochemical reduction occurs. According to Davison (1993) sunlight may affect the reduction and oxidation of Mn in water. The short residence time of water in upland streams and rivers means that the water chemistry is usually not in equilibrium and river pH may not adequately account for in-stream processes (Laxen *et al.* 1984).

Differences in behaviour of iron and manganese are largely attributed to Fe(II) oxidizing more rapidly than Mn(II) and the relative case of reduction of manganese oxyhydroxides.

Lakes and reservoirs

Water in lakes and reservoirs has much longer residence times than in rivers and is more likely to attain a state of chemical equilibrium. Temperature play major role in forming lake ecology and influence the chemical composition of lake water. It results in lake stratification in terms of water temperature, density, oxygen content and as a result, iron and manganese concentrations.

In regions where the temperature in winter drops below 4°C and the lake is deep enough, a seasonal thermal stratification may develop. The water body may have uniform temperature and chemical composition in spring. During summer, as the solar radiation, warms up the top part of water heat up more quickly than the bottom one and three major layers within the lake can be distinguished:

- Epilimnion – an upper layer of circulating warm water, usually no more than 6 m deep, where dissolved oxygen concentrations are moderate to high.

- Thermocline – a layer of rapid temperature and oxygen content decrease with depth, separating the upper and lower layers.
- Hypolimnion – a cold, deep-water, stagnant layer in which oxygen level is low.

In the epilimnion oxidation of Fe(II) and Mn(II) occurs to produce particulate hydrooxidates which sediment into the hypolimnion. On the other hand reducing conditions in the hypolimnion encourage reduction of Fe (III) and Mn(IV) in the water column and bottom sediments to Fe(II) and Mn(II). Those metals and other compounds (sulphur as hydrogen sulphide) may become increasingly soluble and when released from anoxic bottom sediments may cause additionally taste and odor problems.

During late autumn the temperature of the epilimnion drops down and as it approaches or becomes less than that of the lower layer, the density difference and wind forces acting on the surface are powerful enough to cause a reversal of the layers allowing the water from hypolimnion to rise to the surface. In the autumnal turnover of water, dissolved oxygen is more evenly distributed throughout the water body and some reoxidation of Fe(II) and Mn(II) occurs.

Seasonal temperature variations additionally modify the vertical structure and red-ox conditions (Cornett & Rigler 1987). As an effect of this phenomenon iron and manganese concentrations in lake and reservoir waters can vary greatly depending on the season and depth (Boehrer & Schultze 2008).

In shallow lakes, especially those in windy conditions the stratification is unlikely to be a common occurrence since standing waters tend to be well-mixed and iron and manganese concentrations are more uniform.

KEY POINTS:

(1) Iron and manganese are natural elements commonly occurring in the earth's crust, groundwaters and surface waters.
(2) The main factors determining the solubility of iron and manganese and their migration in groundwater are reduction-oxidation potential and pH.
(3) In groundwater, iron usually occurs in the bivalent form, while in surface water – in the form of colloidal suspension of trivalent iron.
(4) Concentrations of iron and manganese in lake and reservoir waters can vary greatly depending on the season and depth.

Chapter 3

Sources of iron and manganese in drinking water distribution network

3.1 GENERAL CONSIDERATIONS

Iron and manganese occurring in the drinking water supplied to consumers may be a result of the natural composition of the water or may be due to secondary contamination. This contamination may take place within the catchment, in treatment or during the flow of the water in the distribution system, both in the water pipe network and in property installations.

In the water distribution system, the source of iron and manganese may be deposits and products of corrosion deposited on walls of pipes or construction materials of which the water distribution system is built (pipes, fittings, tanks).

Iron and manganese deposits found in a water distribution system are formed as a result of two independent processes:

(1) Naturally high concentrations of iron and manganese in water entering the distribution system.
(2) Corrosion processes in the distribution network and internal pipe-work (mainly cast iron, steel and galvanised pipes) as a result of contact with water.

3.2 CONTAMINATION WITH IRON DURING TREATMENT

The technological process in which secondary contamination of water with iron ions occurs is a process of coagulation. As a result, in treated water, excessive concentrations of iron or aluminium cations, from the coagulants used may be formed.

It is observed when the hydrolysis products of coagulants are not fully used for destabilization of the impurities removed from water or when cations of coagulants, with organic impurities, form water soluble complexes. Most often, the reason for this situation is using non-optimal doses of coagulant, inappropriate water reactions, low temperatures and the presence of organic ligands of high affinity to metals (Świderska-Bróż 2001; Sarin *et al.* 2004a; Husband & Boxall, 2011).

Additionally independently of the technological processes, the release of iron ions at the stage of water treatment may occur as a result of corrosion of pipe-lines, installations and elements of equipment of the whole process line of water treatment.

3.3 IRON AND MANGANESE IN WATER DISTRIBUTION SYSTEMS

Corrosion of iron

Iron corrosion is a very complex process. Due to a high variability of conditions in water distribution systems, individual indexes, which determining the corrosion processes, may be critical in one system, and relatively insignificant in the other. Moreover, corrosion itself manifests itself in different ways and can evolve in different ways.

Corrosion may destroy pipes, use up oxidants and disinfectants in water, cause formation of deposits which increase the value of energy needed for delivery of water, promote development of biofilms and cause formation of suspensions composed of iron particles that causes red, brown or yellow discolouration of water (Sarin *et al.* 2004a).

The basic parameters and components of water quality, influencing corrosion processes in water distribution systems are (Sozański & Huck 2007):

(1) Water temperature.
(2) Dissolved oxygen, carbon dioxide, acidic carbonates and calcium carbonates.
(3) Redox potential.
(4) Chlorides and sulphates.
(5) Water flow regime.

In water distribution systems, all three basic corrosion types may occur:

- Chemical corrosion.
- Electrochemical corrosion.
- Microbiological corrosion.

Iron corrosion is related to the release of iron ions from surface of metal:

$$Fe \longrightarrow Fe^{+2} + 2e^{-} \text{ (anode)}$$

Electron acceptors such as oxygen, chlorine or hydrogen ions present in the water enable reaction completion. In water of neutral pH, in the presence of dissolved oxygen, the oxidation reaction is mainly a cathodic reaction:

$$O_2 + 4H^{+} + 4e^{-} \longrightarrow 2H_2O \text{ (cathode)}$$

As effective electron acceptors, hydrated chlorine molecules may also act, such as HOCl and NH_2Cl, when used for water disinfection:

$$HOCl + H^{+} + 2e^{-} \longrightarrow Cl^{-} + H_2\,O \text{ (cathode)}$$
$$NH_2Cl + 2H^{+} + 2e^{-} \longrightarrow Cl^{-} + NH_4{}^{+} \text{ (cathode)}$$

In case of absence of dissolved oxygen and disinfectants, at the appropriate pH, cathodic reactions occur, resulting in the production of gaseous hydrogen due to the reduction of protons:

$$2H^{+} + 2e^{-} \longrightarrow H_2 \text{ (cathode)}$$

Microbiological corrosion is, apart from chemical and electrochemical corrosion, the third basic type of corrosion, in which microorganisms (mainly bacteria) play an active role. Among the bacteria initiating microbiological corrosion, the following should be mentioned: bacteria reducing sulphates to sulphides

(*Desulfovibrio*), bacteria oxidizing sulphur to sulphates with the formation of sulphuric acid (*Thiobacillus*) and bacteria oxidizing iron to iron oxides (*Gallionella* and *Spherotilus*) (Olańczuk-Neyman 1988, 2001; Sozański & Huck 2007).

As a result of corrosion, iron is released into water as dissolved Fe(II) ion. In the presence of oxidants it is oxidized to Fe(III) which forms molecular suspensions of low solubility and at sufficiently high concentrations of iron to impart yellow, orange or brown colour to the water and cause the phenomenon of rust-coloured red water (Sarin *et al.* 2004a). Iron compounds formed as a result of corrosion may be dissolved in water or may be deposited on corroded metal surface. When the deposited iron increases corrosion deposits and decreases corrosion velocity, iron dissolved in water cause equalization of iron solubility.

Formation and structure of corrosion deposits

Water pipe deposits containing iron and manganese found in water distribution system may have two types of origin:

- The result of corrosion processes in the water distribution system.
- The effect of insufficient water treatment (high iron and manganese concentrations in water entering distribution system).

The corrosion rates of metals containing iron has a small relationship with the amount of iron that is actually released into the water. Firstly, corrosion leads to the layering of oxidized iron or other components in the form of deposits which accumulate as large reservoirs of corrosion products (McNeill & Edwards 2000). Iron deposits are typically composed of many different components; therefore they are very heterogeneous (Łomotowski, 2007; McNeill & Edwards 2000; Szuster-Janiaczyk 2011). The most common products of steel corrosion in water environment are presented in Table 3.1.

Table 3.1 The most common products of steel corrosion in water environment.

Compound/mineral	Chemical formula	Density [g/cm^3]	Colour
Iron hydroxide (II)	$Fe(OH)_2$	close to water density	green
Iron hydroxide (III)	$Fe(OH)_3$	1.59	brown
Hydrated iron oxides	$Fe_yO_x(OH)_{3y-2x}$	1.2–1.4	brown
Geothite	α-FeOOH	3.3–4.2 mean 3.8	brown, brown-yellow, yellow ochre
Ferroxide	β-FeOOH	4.2	brown
Lepidocrocite	γ-FeOOH	4.0	red, yellow-brown, blackened brown
Hematite	α-Fe_2O_3	5.3	black, red-grey, blackened brown
Maghemite	γ-Fe_2O_3	4.9	dark brown
Magnetite	Fe_3O_4 ($FeO \cdot Fe_2O_3$)	5.1–5.2 mean 5.15	grey-black, black
Wustite	FeO	5.88	grey
Siderite	$FeCO_3$	3.7	green, yellow-brown

The composition and structure of iron deposits in a water distribution system depends on the type of metal and composition of the aqueous phase. Their formation depends on several factors: parameters of water quality, such as pH, alkalinity, buffer intensity, natural organic matter (NOM) and dissolved oxygen (DO). Moreover, parameters such as: character of flow, seasonal fluctuation of temperature, microbiological activity, method of water treatment, including dosing of corrosion inhibitors, may also influence the formation and growth of corrosion deposits. Therefore it should be expected that corrosion deposits on iron and steel pipes in individual water distribution system will have unique features (Sarin *et al.* 2004a).

Corrosion iron deposits are characterized by layered structure. The base is corrosion foundation, the next layer is a porous core which contains both liquids and solids, then there is a relatively dense shell-like layer which covers the porous core and ensures the structural integrity of the deposits. The last layer is a surface layer, at the water/deposits border, loosely attached to the shell-like layer. Physical and compositional differences have been observed in deposits formed in old iron and steel pipes, exposed to water of various quality parameters. Basic solutions more often form compact deposits than slightly acidic solutions. Because oxidation of Fe(II) in a solution is one of the first factors influencing formation of corrosion deposits, anions, which join dissolved iron may have especially potent influence on deposit composition and may influence the intensity of corrosion (Sarin *et al.* 2004a).

Role of microorganisms

The significant role of microorganisms in the process of the formation of deposits in water pipes is mainly attributed to bacteria, especially iron, manganese and sulphate bacteria. Iron binding bacteria, when occur in water in large amounts, impart a metallic smell and tart taste to it, they change turbidity and the colour of water. Among iron and manganese binding bacteria, various groups are differentiated which accumulate oxidized forms of iron and manganese in various parts of cells. They may be external cytoplasmic processes, mucus pads, sheaths, capsules, surface of cell wall or intracellular organelles.

The deposits of iron oxides and hydroxides impart a characteristic brown-yellow, rust colour to cells or bacterial agglomerates, and manganese deposits – brown-red or green-olive. These bacterial growths can cause water pipe clogging which in consequence may lead to the deterioration of hydraulic conditions in the network; because iron-manganese deposits formed over time by microbiological action become hard, compact and can adhere to pipe walls (see Figure 3.1).

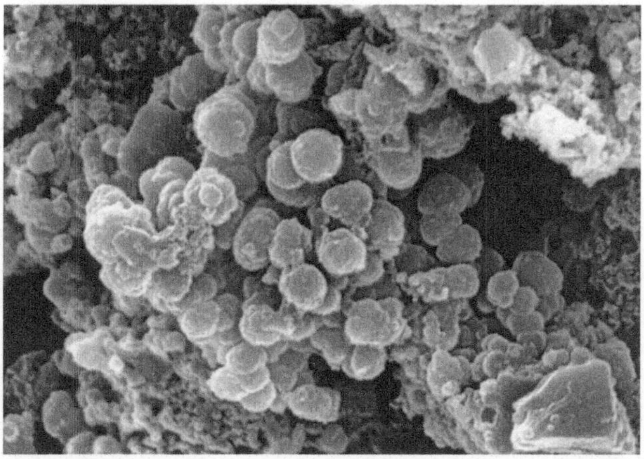

Figure 3.1 Iron deposits of biological origin – magnification 8000 times (Szuster-Janiaczyk 2011).

Bacteria taking part in the formation of corrosion deposits include chemolithotrophic oxidative bacteria – binding iron, but not oxidizing manganese. The most important are: *Acidithiobacillus ferrooxidans*, *Leptospirillum ferroxidans* and species of *Gallionella* genus. The bacteria are able to oxidize Fe(II) under oxygenic conditions, and may use the energy obtained in the process of assimilation of CO_2.

$$FeS_2 + 3.5O_2 + H_2O \longrightarrow FeSO_4 + H_2SO_4$$

However, microbiological reactions of the oxidation of Fe(II) to Fe(III) carried out by *Acidithiobacillus* bacteria proceeds quickly, and the resultant $Fe_2(SO_4)_3$ is a potent oxidant.

$$4FeSO_4 + O_2 + 2H_2SO_4 \longrightarrow 2Fe_2(SO_4)_3 + 2H_2O$$

$$FeS_2 + 2Fe_2(SO_4)_3 \longrightarrow 2FeSO_4 + 2S^0$$

Bacteria of the *Gallionella* genus tolerate pH values from 6.0 to 7.6 in environments in which Fe^{2+} concentration is from 5 to 25 mg/l, oxygen from 0.1 to 1.0 mg/l, and the temperature is from 8 to 16°C. Such conditions are commonly found in groundwater. They oxidize Fe^{2+} to Fe^{3+} according to the reaction:

$$Fe^{2+} + \tfrac{1}{4}O_2 + H^+ \longrightarrow Fe^{3+} + \tfrac{1}{2}H_2O$$

$$Fe^{3+} + 3H_2O \longrightarrow Fe(OH)_3 + 3H^+$$

As a result of metabolic processes, insoluble, brown deposits of iron hydroxide are formed which is accumulated in cells in mucoid pads.

Other bacteria which may participate in the forming of water pipe deposits are the bacteria *Spherotilus – Leptothrix,* which oxidize only Fe or both Fe and Mn. They are arranged as threads surrounded by sheaths in which (or on which) iron oxides are precipitated, and sometimes, manganese oxides.

Bacteria of the family *Siderocapsaceae* with genus *Siderocapsa, Naumanniella, Siderocystis* and, *Siderococcus* (aerobes or microaerophils), are bacteria oxidizing Fe(II) and Mn. This is a group of capsular bacteria not forming threads and pads. Fe and Mn oxides gather on or in capsules (Olańczuk-Neyman 2001). They are found in all places in which iron or manganese as well as organic (organotrophs) substances are found. These bacteria deposit iron compounds in cells, therefore their colour is brown, orange or yellow. Inorganic iron compounds oxidized by iron bacteria proceed according to the reaction:

$$4FeCO_3 + O_2 + 6H_2O \longrightarrow 4Fe(OH)_3 + 4CO_2 + \text{energy (46 kJ/Mol)}$$

Specific groups of bacteria oxidizing iron under anaerobic conditions are anoxygenic phototrophs which use iron and electron donor. Some purple bacteria (e.g. *Chromatinum)* and green (*Chlorobium*) belong to this group.

In the process of anaerobic respiration when the hydrogen acceptor is a mineral compound other than oxygen, the reduction of iron (III) ions to iron ions (II) may occur, described as ferric ion respiration. This process is carried out by, among other things, the bacteria *Alteromonas putrefaciens* (*Shewanella putrefaciens)* which belongs to the Gram-negative bacteria that are able to reduce both iron and manganese.

According to Sikora and Błaszczyk (2009), for determination of microorganisms using Fe(III) as the final electron acceptor in anaerobic respiration, the abbreviation FRM is used (*Fe(III)-respiring microorganisms).* Apart from Fe(III), they reduce Mn(IV) to Mn(II) and humid acids, other metals, radioactive elements, nitrates, fumarate and sulphur. At present, over 100 species of FRM is known, while the best known family is *Geobacteriaceae* oraz *Desulfuromonadaceae* with such genus as *Geobacter* i *Desulfuromonas*.

Reactions on surfaces covered with deposits and their influence on water quality

Water pipe deposits retained in the water distribution system, on the one hand, may protect pipe surface from corrosion processes by limiting oxygen diffusion into metal structure, but on the other hand they may serve as a source of iron and manganese ions contaminating water in a secondary manner. Release of these ions may occur by the way of two interdependent processes. The first process is the tearing off of water pipe deposits due to sudden changes of rate and direction of water flow, or water hammer in the water distribution system. This condition takes place most commonly in cases of damage within the water distribution system (such as a water main burst) and the cessation of water flow related to it. The process is accompanied by migration of iron or manganese to the water in the forms of ions Fe(III), Mn(III). In such cases, concentrations of iron and manganese can exceed regulatory standards and result in an episode of rust-coloured/ red water (iron) and black water (manganese).

Another mechanism leading to a release of iron and manganese ions from water pipe deposits is their dissolution. The process of the release of iron and/or manganese ions as a result of dissolution of water pipe deposits usually is characterized by lower dynamics compared to the process of deposit tearing off. It may be initiated, among other things, as a result of:

(1) Pumping chemically unstable water to a water distribution system.
(2) Change of a source of water supply.
(3) Mixing in a distribution system water from various sources and of various compositions,
(4) Intentional introduction of compounds causing dissolution of iron and manganese deposits into the water (e.g. some corrosion inhibitors based on phosphoric acids).
(5) Reducing conditions in a water distribution system, in which Fe (III) and Mn(III) ions are reduced to Fe(II) and Mn(II) forms.

The occurrence of reducing conditions in a water distribution system may be caused by many factors. It may be as a result of pumping water with low oxygen content into a water distribution system or water of average oxygen concentration, but a significant concentration of ammonium ion. It may also be a result of the action of bacteria reducing iron and manganese ions (Sikora & Błaszczyk 2009). The other method of achieving the reducing conditions in a water distribution system is a release of ammonium ions from water pipe deposits (Lipponen *et al.* 2002; Lipponen *et al.* 2004; Pintar & Slawson 2003). This ion is formed in deposits, among other things, as a result of the activity of denitrifying bacteria (Odell *et al.* 1996; Wilczak *et al.* 1996). The presence of reduction zones in water pipe deposits was shown in many publications (Huang & Zhang 2005). Nawrocki *et al.* (2010) report that, among other things, steady water present in the depth of water pipe deposits, has reductive properties and causes fast reduction of nitrates and formation of unwanted ammonium nitrogen.

It is very difficult to model both mechanisms of a release of iron from water pipe deposits because these deposits are often poorly crystallized and heterogenic. Even when the process of deposit dissolution is understood, the contact of unsaturated water with it may lead to limitations at the stage of kinetics and transfer of mass. In the case of erosion of the deposits, the mechanisms of detachment of single iron molecules depend, among other things, on: resistance and adhesiveness of deposits, velocity of water and, water temperature change.

Factors which may contribute to a release of iron from water pipe deposits are, among other things: corrosion rate, solubility of corrosion deposits, and the oxidation rate of iron ions, which diffuse from deposits. Even though the chemical composition of water inside deposits may differ from the quality of the bulk water, exclusively because of the structure of deposits and their properties, it is understood that changes in the quality of bulk water may also influence the chemical composition of water contained in the pores of deposits and the properties of deposits (Nawrocki *et al.* 2010; Sarin *et al.* 2004a).

The key factors controlling iron release from corrosion deposits to water are: dissolved oxygen, pH value, alkalinity, water temperature, corrosion inhibitors, calcium carbonate, chlorides and sulphates. The influence of an increase of dissolved oxygen concentration on the pipe metal corrosion or on the level of iron release will depend on characteristics of corrosion deposits. In the case of the presence of water pipe deposits in a water distribution system, dissolved oxygen is used both in the corrosion reaction, and as a result of oxidation of iron ions from deposits.

Exposure of deposits to high concentration of dissolved oxygen may reduce iron release by:

(1) Oxidation of already dissolved iron Fe(II) contained in the deposits to less soluble form Fe(III)
(2) Precipitation of Fe(III) phase inside deposits rendering them more dense and less prone to diffusion of Fe(II).

Running water increases the level of transport of oxidants to deposit surfaces and therefore may reduce iron release compared to stable water. Highly porous, still conducting deposits may be less protected from the corrosion of metals covered by them and high concentrations of dissolved oxygen may cause an increase in the amount of released iron (Sarin *et al.* 2004a). The mechanism of corrosion and iron ions release from deposits under aerobic and anaerobic condition are shown in Figure 3.2.

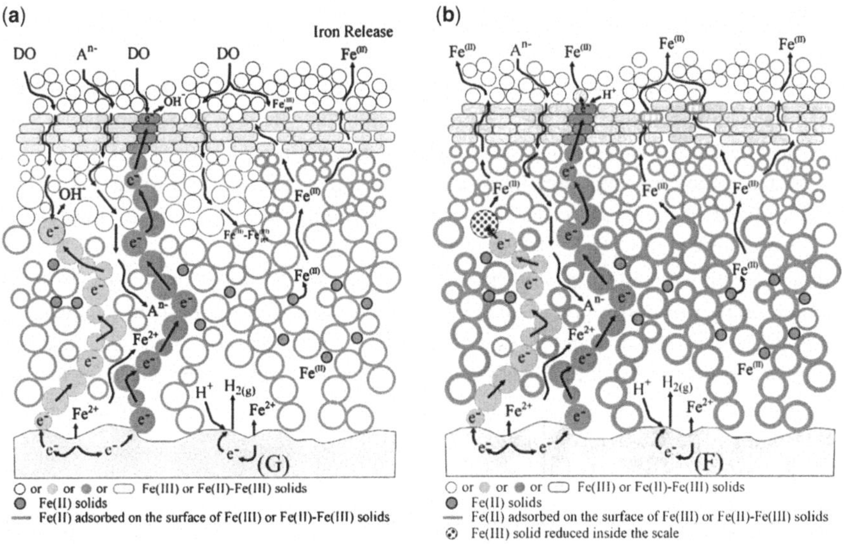

Figure 3.2 Schematic representations of different corrosion reactions and iron release pathways in iron corrosion scales (a) under oxidizing conditions, (b) under anoxic condition (Sarin *et al.* 2004a, with permission).

An increase in the pH of the water promotes formation of iron oxides and hydroxides with lower porosity which are less susceptible to dissolution than the solid iron metal. A decrease in the amount of iron released from water pipe deposits was found with pH increase from 7.5 to 9.5 (Sarin *et al.* 2004a). The density of the structure of the deposits should lower the level of transport of ions through deposits and therefore decrease corrosion intensity and the level of iron release. Calcium carbonate plays a significant role in the process of iron release. Calcite deposits may block pores inside corrosion deposits and therefore hinder migration of iron ions from deposits to water. Chlorides and sulphates play a significant role in the determination of electrical inertness inside the core of deposits and significantly influence the level of iron release from deposits to water (Sarin *et al.* 2004a).

Corrosion of galvanised service pipes and domestic installations

Galvanised steel/iron pipes have often been used as water supply lines for new premises (especially prior to the 1960s) and for domestic pipe-work in many countries. Today galvanised iron is still very often found in water well applications of all types.

During galvanising, steel is immersed in a molten zinc bath, ensuring a tough, uniform barrier coating. Zinc's natural corrosion properties provide long-term protection, in most environments. The zinc reacts with available oxygen in the environment to form zinc oxide, which then forms zinc carbonate after reacting with carbon dioxide.

Galvanised steel can last usually forty to sixty years until it needs to be replaced. Over time many pipes made of galvanized steel have filled with scale, which is why some older homes suffer from low water pressure. Hard water greatly reduces the life of galvanized steel pipes.

In spite of its many benefits, such as mechanical and cathodic protection of base steel, relatively low cost and long lifespan, galvanized steel has also some constraints:

- Galvanised pipe has a tendency to react with minerals in the water, producing a build-up of calcium deposits (scale) that can lead to lower water pressure.
- When mixed with yellow brass, galvanized steel triggers dezincification, and it results in electrolytic action when combined with nonferrous metals, such as copper and brass. The effects of galvanic corrosion are usually limited to the immediate area of metal-to-metal contact (St. Clair *et al.* 2012). In most cases the use of dielectric fittings helps minimise the problem.
- Galvanised steel pipes may contain lead, which corrodes quickly and reduces the lifespan of the piping. Galvanized iron pipes can release significant amounts of lead into standing water, as the zinc coating contains about 1% lead impurities (AWWARF 1996).
- Homes that have galvanized pipes and have or had lead service lines are at risk from the release of lead in water from corroded pipes. Galvanised piping may accumulate lead deposits over time. As galvanized pipes corrode they form rust, lead that is accumulated over decades is likely to be found deep in the interior walls of rusty pipes. Galvanised pipes may continue to serve as a lead source in drinking water long after all other sources of lead have been removed (Giani 2008; HDR 2009; Deshommes *et al.* 2010; McFadden *et al.* 2011).

3.4 SECONDARY WATER CONTAMINATION BY IRON AND MANGANESE

The processes of secondary water contamination, among other things, with iron and manganese ions, should be prevented as early as possible at the stage of planning of water intakes, water treatment stations and water distribution system so that the technologies used, materials, hydraulic systems or the method of their management do not favour processes of corrosion and accumulation of water pipe deposits (DHS 1992; Dodril & Edwards 1995; Grossblatt, 1998; Husband & Boxall 2011; Szuster 2003; Szuster-Janiaczyk 2004, 2005).

Water treatment and corrosion inhibitors

In order to prevent processes of secondary water contamination in the distribution network with iron and manganese ions, first of all, it should be ensured that water supplied to the system is chemically and biologically stable. Also, mixing water originating from various delivery sources of different composition in water distribution system should be prevented. Even when source waters have stable water qualities, the final water quality may vary due to varying proportions of mixing components (DVGW 2004). German guidelines indicate that in the analysis of stability of mixing waters in water distribution system the following parameters should be considered: water temperature, dissolved oxygen (DO), pH, mineral acidity to pH = 4.3, chlorides, nitrates, phosphates, dissolved organic carbon (DOC), sodium, potassium, calcium and magnesium.

Many publications suggest that the central use of corrosion inhibitors, among other things, prevent excessive corrosion. However, it should be remembered that the effects of corrosion inhibitors usage may vary depending on distribution system and water quality. Each particular system should be thoroughly investigated before full-scale usage of inhibitors. Even the theory of preventive anticorrosion action of phosphates itself changes significantly over time. The studies on phosphates corrosion control do not always find beneficial effects (Edwards 2004; McNeill & Edwards 2000, 2002; Sarin *et al.* 2004a, 2004b; Zhang & Edwards 2012).

It was not sufficiently explained, if in the process of corrosion control, ortho- or polyphosphates play the predominant role. According to some theories, ortho-phosphates PO_4^{3-} are extremely stable corrosion inhibitors and they form stable, phosphate protective layers (AWWARF 1990). Other theories explain that polyphosphates are adsorbed on iron surface aiming to form a protective layer, whereas the significance of influence of calcium presence on polyphosphates efficiency is important. It may allow formation of thin calcite layers which gain their protective action by adsorption of polyphosphate ions. In the same paper, a significance of flow conditions is stressed, conditioning efficacy of polyphosphate action (McNeill *et al.* 2000).

In the literature, anticorrosion action is attributed to orthophosphates or polyphosphates. However, this division is not clear, because many polyphosphate corrosion inhibitors undergo reversion in the network and act further as orthophosphates.

According to most commonly stated theories, orthophosphate inhibitors containing phosphoric acid (H_3PO_4) and its salts of alkali metals (Li, Na, K, etc.) are non-oxidizing, buffering, anode inhibitors in the corrosion reaction, and an inhibitory mechanism is based on the inhibition of corrosion processes, proceeding according to the equations:

$$Fe_{(metal)} + 0,5O_2 + H_2O \longleftrightarrow Fe^{+2} + 2OH^-$$

(under condition of oxygen-running water access)

$$Fe_{(metal)} + 2FeOOH_{(deposit)} + 2H^+ \longleftrightarrow 3Fe^{+2} + 4OH^-$$

(under conditions of oxygen deficit-stagnating water)

According to Sarin (*et al.* 2004a), the role of orthophosphates as anode inhibitors is secondary, and corrosion control by them is obtained, most probably, by a control iron release, as a product of corrosion. According to this theory, orthophosphate dosing leads to a decrease of porosity of water pipe deposits. It is made by precipitation of iron phosphates (e.g. $FePO_4 \cdot 2H_2O$) which form some kind of "plugs" in the pores, on the surface layer of deposits of iron oxides or in its deeper layers. It results in their lower porosity, thus making it more difficult for the diffusion of Fe^{+2} ions, being the product of iron corrosion relocating outside and for diffusion of anions migrating from water to the core of corrosion deposits. The high penetration of phosphate ions into the space of deposits, as occurs under conditions of flow, should facilitate formation of protective "stoppers". It has not been possible to explain the influence of other corrosion inhibitors on a decrease of iron release from corrosion deposits (Sarin *et al.* 1995, 2001, 2004a, 2004b).

It was reported in many papers that polyphosphates may prevent corrosion and/or control the phenomenon of formation of rust-coloured water, but prevention of corrosion and formation of rust-coloured/red water are in fact two different phenomena (Hater *et al.* 2002; McNeill & Edwards 2000).

According to the American Chemistry Council, polyphosphates, alone or combined with orthophosphates, may effectively control corrosion of iron metals and non-iron metals and their alloys, and polyphosphates alone are especially effective as factors used for corrosion control in lower temperatures and at pH reaction lower than 7.5 (ACC 2004).

Among publications concerning corrosion control with the use of FIK, are also publications which indicate the lack of their obvious beneficial effect. A four-year study conducted by McNeill and co-workers (McNeill & Edwards 2000) in which under changing conditions, various corrosion inhibitors were studied (polyphosphate $Na_{22}P_{20}O_{61}$, orthophosphate Na_3PO_4, zinc orthophosphate $[Zn_3(PO_4)_2]$) led to a conclusion that addition of phosphate corrosion inhibitors had no statistically significant influence (95% level of confidence) on the amount of iron released in corrosion processes. In several studies carried out in the frame of experimental studies, addition of orthophosphates or zinc orthophosphate caused an increase of iron concentration. Addition of polyphosphates decreased iron release, only in one case and caused an increase in other cases. Dosing of orthophosphates reduced total loss of mass in cases of pipe corrosion only in cases of water of one quality in all other cases, addition of orthophosphates caused the opposite effect (Edwards 2000; McNeill & Edwards 2000).

According to Edwards (2000), addition of polyphosphates or orthophosphates under conditions of long water stagnation, rarely caused any benefits in the process of corrosion control. In cases when statistically significant effects were seen, inhibitors tended rather to an increase of iron concentration increase and general decrease of sample mass compared to pipes to which no phosphates were dosed.

Protection against excessive growth of iron bacteria in the water distribution system is based primarily on proper oxidation of the water during aeration process in water treatment plant. In this process ferrous compounds become oxidized. Depriving water of these ferrous compounds, bacteria are deprived of the source of energy at the same, which is oxidation of divalent iron (Fe^{2+}) to trivalent hydroxides ($Fe(OH)_3$). The presence of iron in water and water stagnation (mainly in ducts and tanks) favours development of iron bacteria which will accumulate deposits of iron hydroxide in their environment.

Protection from manganese bacteria is more complex. It can be carried out at treatment works in which sand filters are used with inoculated manganese bacteria. These bacteria, for their metabolic processes, capture manganese from water, due to this it is used by these bacteria and manganese concentrations in water decrease so that in further sections of a network, manganese bacteria have not the conditions for development. In case when filters do not capture manganese, chemical precipitation of manganese may be used.

Exploitation regime and distribution network arrangement

The intensity of iron and manganese ions release is strictly related to the hydraulic regimen in the water distribution system, the quality of materials used for building the water distribution system and their technical condition.

Among hydraulic parameters which should be analysed in the context of secondary water pollution, the most important are: velocity of running water, time of water retention in tanks, time of water stagnation in the water distribution system, pressure in the network and existing mixing zones in network. These factors are conditioned mainly by the arrangement and the method of exploitation of the water distribution system.

Velocity of flow and time of water stagnation in the network depend directly upon the diameter of water pipes. Together with an increase in the diameter of pipes, at constant flow, velocities decrease, and the time of water maintenance in the network increases. In re-dimensioned networks, low velocities and long water stagnation times influence final water quality. In areas of long-lasting water stagnation (network ends, dead points) corrosion products come into water more intensively as well as compounds and element of materials rinsed from mineral deposits and the biofilm.

An especially important factor conditioning water quality in a water distribution system is also the method of network exploitation, taking into consideration, among other things, maintaining constant, pre-defined water pressure. Pressure changes, especially when abrupt, cause sudden changes of velocity and directions of flow, and can cause breaking of loose minerals and corrosive deposits and the biofilm, resulting usually in a sudden deterioration of water quality parameters.

Technical condition of water distribution system

The materials used in the construction of a water distribution system and their quality are significant factors influencing the intensity of iron and manganese ions release in water distribution.

Technical condition of the network is most commonly related to the age of the network and the method of its exploitation. Water pipes, pumps, tanks, valves should operate efficiently. Technical inspections and maintenance procedures should be carried out systematically. These activities should prevent failures, occurrence of which significantly influences quality of water in water distribution system. In order to prevent failures and effects of corrosion, lack of stability of water pumping to water distribution system, activities related to maintenance of proper technical state of the network should be carried out systematically, that is, flushing, cleaning, renovation and exchange of water pipe network.

Network cleaning/flushing

A program of network cleaning is implemented most commonly in order to remove more compact corrosion products and deposits.

Methods commonly used for pipe network cleaning are (Kolonko 1998):

- Mechanical cleaning.
- Hydraulic cleaning.
- Hydro-mechanical cleaning.
- Chemical cleaning.

According to Denczew and Królikowski (2002) flushing of the network may be performed as:

(1) Planned procedures carried out according to specific schedule (determination of the site, frequency of flushing).
(2) Interventional flushing (in response to consumers' complaints).
(3) Flushing carried out in situations which could influence negatively the water quality, such as failures, repairs, rehabilitation.

Through properly conducted, systematic flushing of water pipe network (especially its endings and areas of water stagnation), it is possible to obtain satisfactory results for removal of loose deposits, biofilms and products of corrosion responsible for deterioration of water quality in the network. In over-dimensioned networks, the problem may however, be achieving sufficient velocities of flushing 1.0–1.2 m/s. The solution may be the flushing of the network with the use of a directed flows method, if possible, supported by data derived from hydraulic modelling of the flushed network (Kulbik 2002; Maj 1999) or the use of a method of hydro-pneumatic flushing with water and air (air scouring). However, the drawback of the last method is that during flushing with the use of the air in the network, vibrations which have an influence on the integrity of the network concerning leakage (Zimoch *et al.* 2004).

When the technical condition of the network is very poor (leakage, numerous compact corrosion products, mineral deposits and biofilm) so that systematically performed flushing and cleaning are not sufficient to maintain appropriate water quality in the network, the solution to the problem may be a renovation of the water pipe network. Criteria deciding about the selection of mains and service pipes for renovation are:

- Intensity of internal deposition causing aesthetic problems.
- Function and role of the pipe in the system.
- Pipe material.
- Age of the pipe.
- Number of pipe failures/breaks per 1 km per year.
- Relative flow in case of thoroughfare.
- Flow velocity.

Moreover, prior to network renovation, socio-economic criteria should also be taken into consideration such as: duration of the process, durability of the solution used, localization of water pipe-centre or outskirts of the city and possible communication problems related to it (Kolonko 1998).

The final solution which may radically improve water quality is replacement of pipes (with or without excavation) targeting the oldest, most affected sections of the network or fragments made of a material being a source of additional impurities. Similarly, as in the case of renovations and repairs of the network, before exchange of the network sections, criteria are needed for deciding on the selection of a pipe for exchange and for selection of a method (Kolonko 1998).

KEY POINTS:

(1) Iron and manganese in distributed water may come from natural raw water or from the distribution network.
(2) Iron and manganese deposits found in water distribution systems are formed as a result of pumping into the distribution system water containing high concentrations of these elements or the corrosion of the materials (mainly steel and cast iron) in contact with water.
(3) The technological process in which secondary contamination of water with iron ions occurs is the process of coagulation.
(4) The intensity of iron and manganese release is strictly related to the hydraulic conditions in the water distribution system, quality of materials used for building the water distribution system and their technical condition.
(5) The significant role in deposits formation is attributed to bacteria, especially iron, manganese and sulphate bacteria.
(6) In order to prevent processes of secondary water pollution in the network with iron and manganese ions, first of all, it should be taken care that water supplied to the water distribution system is chemically and biologically stable.
(7) The importance of corrosion inhibitors in corrosion control, in the water distribution system is significant, but decision about application this kind of treatment, requires detailed research.
(8) Appropriate network management, cleaning and flushing the pipes and tanks, good quality materials, and so on, allow limiting the amount of iron ions going to the taps.

Chapter 4

Regulatory background

4.1 WORLD HEALTH ORGANIZATION (WHO) GUIDELINES

The World Health Organization (WHO) has published guidelines periodically since 1958, first as "International Standards for Drinking Water" (WHO 1958) and later in 1984 as "Guidelines for Drinking-water Quality" (WHO 1984). These were subsequently revised in 2004 (WHO 2004) and again in 2011 when the fourth edition was published (WHO 2011a).

The guidelines are intended to support the development and implementation of risk management strategies that will ensure the safety of drinking-water supplies. That should be achieved through the control of hazardous constituents of water. The guidelines describe minimum requirements for safe practice to protect the health of consumers. The WHO recommend that if a Guideline value is exceeded then the appropriate Health Authorities should be consulted and if necessary, a plan of action for water quality improvement drawn up.

Numerical "guideline values" for constituents of water or indicators of water quality are in general derived on the base of toxicity of the particular constituents but acceptability aspects such as colour or taste are considered. A guideline value normally represents the concentration of a constituent that does not result in any significant risk to health over a lifetime of consumption (WHO 2011a).

A health-based value (0.4 mg/l) was derived for manganese based on the upper range value of manganese intake of 11 mg/day, identified using dietary surveys, at which there were no observed adverse effects. However, as this health-based value is well above the concentrations of manganese normally found in drinking-water, it is not considered necessary to derive a formal guideline value.

Iron in drinking water is present in concentrations that do not present a hazard to health. The taste and appearance of drinking water will usually be affected below this level. No health based guideline value for iron in drinking water is set, however, an acceptability based guideline value was proposed (Table 4.1).

Table 4.1 WHO Guideline values for iron and manganese in drinking water (WHO 2011a).

Metal	Guideline value mg/l	Basis
Manganese	0.4	Toxicity
	0.1	Prevention of dirty water
Iron	0.3	Prevention of dirty water

Most of the national water quality standards worldwide are developed on the basis of WHO guidelines, however, due to local or national environmental, social, economic and cultural conditions, parametric values may vary.

4.2 EUROPEAN COMMUNITY

Directive 98/83/EC

The Council Directive 98/83/EC of 3 November 1998 on the quality of water intended for human consumption sets out standards for parameters, and sampling and monitoring programs to be used by member states. These have to be incorporated into the legislation of member states. Member states may add to the programs but not subtract from them. In practice most non-EU European states have also adopted the principles of the Directive. The standards for parameters (parametric values) are generally the same as the WHO guideline values and rely on the same experimental evidence.

According to the Directive the parametric values are to be complied with at the point where water intended for human consumption is made available to the appropriate user. This is usually a kitchen tap but in many countries (e.g. Poland) it is interpreted as a point that water enters a building/premise or at the main water meter in public or apartment buildings. The value applies to a sample of water intended for human consumption obtained by an adequate sampling method as to be representative of a weekly average value ingested by consumers.

An important difference between the WHO guideline values and the EU parametric values is that whereas the guideline values are expressed as annual means, the Directive parametric values are expressed as maxima, that is, never to be exceeded.

The Directive 98/83/EC also sets out "Indicator parameters" which member states may wish to adopt as a measure of good practice but are not obliged to do so. The indicator parameters that are shown in Table 4.2 are mostly based on reduction of dirty water in distribution. That particularly applies to iron and manganese.

Table 4.2 EU Directive 98/83/EC Annex 1 part C – indicator parameters, parametric values for metals.

Metal	Parametric value
Aluminium	200 µg/l
Iron	**200 µg/l**
Manganese	**50 µg/l**
Sodium	200 mg/l

The Directive also sets out minimum frequencies of sampling based on the significance of the parametric values and the size of population served (see Chapter 7). The sampling frequencies for many of the parameters are sufficiently low to mean that many years' data is needed to judge real compliance or failure with 95% confidence where the measured values are close to the parametric values. The performance characteristics of the analytical methods to be used for the monitoring of metals are specified, member states are free to use any method which complies with the specified performance parameters (trueness, precision and limit of detection).

Directive 75/440/EEC

The Council Directive 75/440/EEC of 16 June 1975 concerning the quality required of surface water intended for the abstraction of drinking water in the Member States is a historical standard for raw waters intended for production of drinking water. The purpose of this Directive was to give guidance to the member states on the degree of water treatment needed to produce satisfactory drinking water from various qualities of raw water and to ensure that raw water quality did not deteriorate. Three classes of raw surface water: DW1, DW2 and DW3 are defined by reference to water quality parameters (Table 4.3).

Table 4.3 Surface water classifications according to Council Directive 75/440/EEC (selected parameters).

Parameter	Units	Classification					
		Category A1		Category A2		Category A3	
		G	I	G	I	G	I
Dissolved iron	mg/l	0.1	0.3	1	2	–	1
Manganese	mg/l	0.05	–	0.1	–	1	–

G – guide.
I – mandatory.

Several metals are included in the Directive and relate to the ability of the processes commonly used in drinking water to reduce the metal concentrations to below the maxima recommended by WHO.

For each class of water a minimum treatment requirement is laid down in Annex I to the Directive (Table 4.4). Surface water which does not fall within one of these classes cannot be abstracted for supply except by special consent.

Table 4.4 The standard methods of treatment for transforming surface water of categories A1, A2 and A3 into drinking water.

Category	Treatment methods
A1	Simple physical treatment and disinfection, for example, rapid filtration and disinfection.
A2	Normal physical treatment, chemical treatment and disinfection, for example, pre-chlorination, coagulation, floccularion, decantation, filtration, disinfection (final chlorination).
A3	Intensive physical and chemical treatment, extended treatment and disinfection for example, chlorination to break-point, coagulation, flocculation, decantation, filtration, adsorption (activated carbon), disinfection (ozone, final chlorination).

The directive 75/440/EEC has been superseded by parts of the Water Framework Directive (2000/60/EC). However, in many European water companies, standards set by this directive are still recognised as guideline values.

4.3 UNITED STATES OF AMERICA

The Safe Drinking Water Act (SDWA) is the main federal law that ensures the quality of drinking water in the United States of America (US). It was first published in 1974 and it was amended in 1986 and 1996.

The SDWA authorizes the Environmental Protection Agency (EPA) to establish minimum standards to protect tap water and requires all owners and operators of public water systems to comply with health-related standards. The EPA sets standards for drinking water as "National Primary Drinking Water Regulations" (NPDWRs) or primary standards. These standards are legally enforceable and apply to public water systems. The standards tend to be similar to the WHO guideline values as they are largely based on the same experimental evidence.

Primary standards are to protect public health by limiting the levels of contaminants in drinking water. In the US, this is achieved by periodic surveys at consumers' taps with sample numbers reflecting the population of the water supply system being assessed.

Iron and manganese are not subjects of NPDWRs. These metals are considered as constituents that are not toxic but may cause acceptability problems and are included in "National Secondary Drinking Water Regulations" (NSDWR), also called "secondary standards".

NSDWRs are non-enforceable guidelines regulating contaminants that may cause cosmetic effects (e.g. skin or tooth discoloration) or aesthetic effects (odour, or colour) in drinking water. The US EPA recommends secondary standards to water systems but does not require systems to comply. Individual states may however, choose to adopt them as enforceable standards. Maximum contaminant levels (MCLs) for public water systems set by the National Secondary Drinking Water Regulations, part 143, are presented in Table 4.5. The US EPA specifies minimum sampling frequencies. Under the NSDWA, the US EPA also establishes minimum standards for state programs to protect underground sources of drinking water.

Table 4.5 Secondary maximum contaminant levels for iron and manganese (SMCL) according to US EPA.

Metal	Level
Iron	0.3 mg/l
Manganese	0.05 mg/l

4.4 OTHER COUNTRIES

In Canada, the "Guidelines for Canadian Drinking Water Quality" and the "Guideline Technical Documents" are developed by the Federal-Provincial-Territorial Committee on Drinking Water and published by Heath Canada (www.healthcanada.gc.ca/waterquality).

Australia and many other countries around the world have set their own regulatory standards. These tend to follow WHO guideline values with additions for local water quality problems. Standards for iron and manganese in drinking water in selected countries are presented in Table 4.6.

Table 4.6 Standards for iron and manganese levels in drinking water in selected countries.

Country	Standard value in mg/l	
	Iron	**Manganese**
Canada	0.3(A)	0.05(A)
Australia	0.3(A)	0.05(H); 0.01(A)
Japan	0.3	0.05
South Africa	0.2(A) (Cl. I); 2(Cl. II)	0.1(A) (Cl. I); 1(Cl. II)
Taiwan	0.3(A)	0.05(A)
India	0.3	0.1

A – aesthetic.
H – health based.
O – operational.
Cl. I – Class I (recommended operational limit).
Cl. II – Class II (max. allowable for limited duration), water consumption period max. 7 years.

KEY POINTS:

(1) The WHO guidelines for drinking water quality are intended to support the development and implementation of risk management strategies that will ensure the safety of drinking-water supplies.
(2) WHO set no health based guideline values for iron and manganese in drinking water. The concentrations of these metals normally found in drinking-water do not present a hazard to human health.
(3) Standards for iron are set for acceptability aspects.
(4) Standards for manganese are set for both health and acceptability aspects. Standards set for aesthetic aspects are usually more stringent.

Chapter 5

Health perspectives

5.1 MANGANESE

Manganese as an essential nutrient

Manganese is one of the most abundant metals in soil and it occurs as oxides and hydroxides. Manganese is an essential element for all species (animals and vegetables). Some organisms, such as diatoms, molluscs and sponges, accumulate high amounts of manganese. Fish can have up to 5 ppm and mammals up to 3 ppm in their tissue, although normally they have around 1 ppm (US EPA 2004).

Manganese is an essential mineral nutrient needed for the proper functioning of the human body. Manganese is a constituent of several metalloenzymes and acts as an enzyme co-factor (i.e. superoxide dismutase, glutamine synthetase, arginase, but also oxidoreductases, transferases, hydrolases, lyases, isomerases, and ligases). It is an important compound for enzymes involved in carbohydrate, amino acids, proteins, lipid and sterol metabolism, and in oxidative phosphorylation generating cellular energy-ATP (Fraga 2005; WHO 2002).

Manganese is involved in the function of numerous organ systems and it is needed for normal immune function, for blood sugar regulation, for production of cellular energy, reproduction, digestion, and bone growth. Manganese also links with vitamin K to support clotting of the blood. As a component of superoxide dismutase (SOD), manganese is an important species with antioxidant properties. Indeed, manganese-SOD (Mn-SOD) is one of the body's main front-line defence mechanisms against damaging free radicals (WHO 2011a, b).

The main source of manganese in the general population is from food. Foods like nuts, grains, fruits, legumes, tea, leafy vegetables, infant formulas, and some meat and fish are rich in manganese (ATSDR 2000; IOM 2002; US EPA 2004). Daily consumption of manganese originating from food range between 0.7 to 10.9 mg/day and can be even higher among vegetarians and heavy tea drinkers (ATSDR 2000; Freeland-Graves *et al.* 1987; Greger 1999).

Manganese is also an essential mineral nutrient needed for proper foetal development. The findings of two recent studies (Vigeh *et al.* 2008; Zota *et al.* 2009) indicate that lower maternal blood manganese is associated with foetal intrauterine growth retardation (IUGR) and lower birth weight (Wood 2009). Today, manganese deficiency in humans appears to be rare because it is present in many common foods. Manganese intake deficiency also cause seizure activity, leads to very poor bone formation, impairs fertility, and causes birth defects in humans, but also can manifest as transient dermatitis, hypocholesterolemia, and an increased alkaline phosphatase levels in liver (US EPA 2004; WHO 2011a, b). The loss of these cells results in reduction of a dopamine, the chemical messenger responsible for coordinated muscle movement.

There are differences in manganese intake between breast-fed and formula-fed infants up to age 6 month due to different concentrations of manganese in human milk and infant formula. Human milk contains on average 3.5 to 7.5 μg/l manganese and infant formula 100 times higher concentrations of this metal (Ljung & Vahter 2007). Consequently, the similar ratio exists in estimated daily weight-adjusted manganese intake in infants assuming an average weight of 6 kg for an infant age 6 months and intake of 780 ml of milk per day (IOM 2001). After the introduction of solid foods, contribution of manganese intake from milk is decreased (US EPA 1996). Manganese concentration in drinking water which the formula is mixed should be taken into consideration when the infant's daily manganese exposure is assessed.

According to the Nutrition Board of the Institute of Medicine an adequate intake (AI) value for Mn is 3 μg/l for infants 0–6 months old, which is at the lower end of ingested Mn/day from human milk. For adult men and women the AI is 2.3 mg Mn/day and 1.8 mg Mn/day, respectively. Due to the presence of manganese in foods and drinking water, manganese deficiency in humans is rare (US EPA 2004).

Manganese toxicity

The toxic effects of manganese depend on its exposure level, route of exposure, chemical form, the age, nutritional, health status and inter individual variation. Inhaled manganese may cause toxic effect in much lower doses than after oral ingestion due to the retrograde transport from the olfactory epithelium to the striatum of the brain (Gianutsos *et al.* 1997). In the case of the oral route of exposure manganese is first absorbed from the gut in divalent form and then reaches the blood and is distributed through the body.

The total percentage of dietary manganese absorbed from a meal is small; most studies suggest that humans absorb <5%. The uptake is regulated so that when dietary manganese levels are high, the gastrointestinal absorption is reduced (Ljung & Vahter 2007). The factors that influence Mn absorption include the carbohydrate source in the diet, the presence of phytate, the presence of animal protein, the manganese content of the diet, and the dietary content of other mineral elements, especially iron (Finley 1999). The main route of manganese excretion is through the biliary system. Infants and young children are more susceptible to high manganese levels due to the incompletely developed biliary system, greater retention and higher affinity of certain tissue sites for manganese than in adults (Lonnerdal 1994). A small part of absorbed manganese is excreted by urine, sweat, hair and human milk (US EPA 2004).

Manganese effects occur mainly in the respiratory tract and in the brain. Symptoms of manganese poisoning are hallucinations, forgetfulness and nerve damage. Manganese can also cause Parkinsons disease, lung embolism and bronchitis. When men are exposed to manganese for a longer period of time they may become impotent. A syndrome that is caused by manganese has symptoms such as schizophrenia, dullness, weak muscles, headaches and insomnia. Because manganese is an essential element for human health shortages of manganese can also cause health effects such as: fatness, glucose intolerance, blood clotting, skin problems, lowered cholesterol levels, bone disorders, birth defects, changes of hair colour and neurological symptoms.

People may also be exposed to manganese if they live in high automotive traffic areas where manganese is used as a gasoline additive. Results from some studies revealed that people who inhaled manganese from automobile emissions reflect these exposures with higher levels of manganese in their blood, indeed the people showed signs of neurological problems that were similar to those reported in occupationally exposed individuals (NIEHS 2010).

Occupational studies of very high levels of inhaled manganese compound have shown sufficient evidence for the occurrence of the syndrome known as "manganism", which is a set of neurological signs and symptoms which are very similar to Parkinson's disease (PD). It is characterized by weakness, anorexia, and muscle pain, apathy, and slow speech, monotonous tone of voice, emotionless "mask-like" facial expression, and slow clumsy movement of the limbs. Manganese also affects lung, liver and cardiovascular system and induce reproductive and foetal toxicity (US EPA 2004).

Children are more vulnerable to adverse health effects of manganese because of their sensitive nervous system, their low bile excretion, high gastrointestinal absorption and homogenous diet. Adverse effects are manifested primary as behavioural disorders and intellectual impairment. At an average 800 μg Mn/l in drinking water, a study in Bangladesh found an association between manganese exposure and neurotoxic effects and a dose-response association between water manganese concentrations and test scores of performance as well as verbal ability (Wasserman *et al.* 2006). Woolf *et al.* (2002) reported that a 10-year old boy with abnormal verbal and visual memory function had elevated serum (0.90 μg/dl vs. normal value of $<$ 0.265 μg/dl), whole blood, urine, and hair manganese concentrations following chronic ingestion of well water containing moderately elevated levels (~1.2 ppm) of manganese (Manganese Health Research Program, USA). Differences in behaviour (hyperactivity, oppositional behaviour) were shown in a study that included two groups of children with average 600 μg/l and 160 μg/l manganese in the tap water, respectively (Bouchard *et al.* 2007). Recently, a cross-sectional study which involved 362 children, 6–13 years of age in southern Quebec, Canada reported a 10-fold increase in water manganese to be associated with a decrease of 2.4 IQ points at manganese levels common in groundwater drinking water sources (median Mn level 34 μg/l) and below current drinking water standards of 50 μg/l (Bouchard *et al.* 2011). Nevertheless, more new studies will be needed to confirm these findings (Chen & Copes 2011).

Experimental animal data, especially rodent data, are not appropriate for human risk assessment because the physiological requirements for manganese vary among different species (WHO 2011a, b).

The mechanism for the neurodegenerative damage specific to select brain regions is not clearly understood. Manganese neurotoxicity may be associated with its interaction with other essential trace elements, including iron, zinc, copper and aluminium. The possible mechanism of manganese toxicity can be disturbance in iron homeostasis and the valence state of Mn. Studies in cultures cells have shown increases in cellular iron uptake when cells were exposed to manganese compounds.

On the other hand, recent evidence from human and non-human primate studies provide strong support to the hypothesis that excess levels of manganese accumulation in the brain results in an atypical form of Parkinsonism which is not associated with the degeneration of nigrostriatal dopaminergic neurons as is the case in Parkinson's disease. These new findings suggest that the Mn-induced movement abnormalities are not consequences of the decreased synthesis or concentration of dopamine, but rather a problem of the ability to release the available dopamine. Further investigations are needed to confirm this new suggested mechanism of manganese neurotoxicity (Guilarte 2010).

In order to assess the dose that an individual has been exposed to manganese and the possible risks of negative effects from excessive intake, biomarkers of exposure are used. For this purpose various biological materials like urine, blood, erythrocyte, hair, toenails, and tooth enamel have been used. Generally, urine is not a good indicator of Mn exposure because only small amounts of daily absorbed Mn (1%) can be measured this way (Smargiassi & Mutti 1999). There are numerous limitations in assessing long term manganese exposure when analysing the whole blood, plasma, serum or erythrocyte because of the strong association between iron and manganese. A recently conducted study has indicated toenail as a valid measure of cumulative manganese exposure 7 to 12 months earlier (Laohaudomchok *et al.* 2011), but manganese concentrations in hair appeared to be associated with intake from water, but not with intake from food (WHO 2011b).

In the EC directive on the quality of water intended for human consumption (98/83/EC) the value for manganese is set at 50 µg/l and represents an indicative value that is not legally binding. The Member States concerned must consider whether non-compliance poses any risk to human health and if remedial action to restore the quality of the water is necessary.

The US EPA (2004) determined health advisories (HAs) for one-day, ten-day and life time exposure for manganese from drinking water. Thus, one-day HA and ten-day HA for children and adults are equal and is 1 mg/l. For infants younger than 6 months, the lifetime HA of 0.3 mg/l is advised to be used for an acute exposure of 10 days, because of the concerns for differences in manganese content in human milk and formula and the possibility of a higher absorption and lower excretion in young infants. Lifetime health advisories are only developed for chemicals that are not likely to be carcinogenic to humans. The Lifetime HA represents that portion of an individual's total exposure that is attributed to drinking water and is considered protective of noncarcinogenic adverse health effects over a lifetime exposure. The reference dose (RfD) of 0.14 mg Mn/kg-day (US EPA 1997) is estimated to be an intake for the general population that is not associated with adverse health effects; this is not meant to imply that intakes above the reference dose are necessarily associated with toxicity (US EPA 2004).

In previous editions of the "WHO Guidelines for Drinking-water Quality" the health-based guideline value was 400 µg Mn/l and tolerable daily intake (TDI) was 60 µg Mn/kg body weight (WHO 2004b; WHO 2006). In the recent edition, WHO published that "this health-based value is well above concentrations of manganese normally found in drinking-water; it is not considered necessary to derive a formal guideline value" (WHO 2011a).

5.2 IRON

Iron as an essential nutrient

Iron is an integral part of numerous essential proteins, such as the red colouring agent of the blood haem-containing proteins (e.g. haemoglobin, myoglobin, electron transport chain and microsomal electron transport proteins, catalase), iron-sulphur proteins (e.g. aconitase, fumarate reductase), Fe-activated enzymes such as NADH dehydrogenase, succinate dehydrogenase, alcohol dehydrogenase, cyclooxygenases and proteins for Fe storage and transport (transferrin, lactoferrin, ferritin, hemosiderin) (Arredondo & Nunez 2005; Fraga 2005). Thus, iron is involved in crucial processes in the body: oxygen transfer from lungs to tissues and trans-membrane electron transfer. It is also important for adequate immune response and neuronal functions, which means appropriate activity and numbers of immune cells (macrophages, neutophils and T-lymphocyte), and early life myelination (Arredondo & Nunez 2005).

The main source of iron in the general population is from food. Sources of haem Fe (15% of consumption) are haemoglobin and myoglobin from animals. Sources of non-haem Fe are cereals, seeds of leguminous plants, fruits, vegetables, and dairy products (Fraga 2005). The daily consumption of iron originating from food ranges from 10 to 14 mg (WHO 2003). Drinking-water containing 0.3 mg/l would contribute about 0.6 mg to the daily intake, representing only about 2% to 8% of recommended dietary allowance (RDA) for iron.

Cellular iron homeostasis is regulated by the transferrin receptor, involved in plasma-to-cell Fe transport, and the iron-storage protein ferritin. It is achieved by iron regulated intestinal absorption. Two kinds of signals have been proposed to regulate Fe absorption: the "erythropoietic regulator" and the "stores regulator". The concept of the "erythropoietic regulator" emerged from the observation that intestinal Fe absorption increases when erythropoietic needs for Fe are not satisfied. Similarly, the concept of "stores regulator" emerged from the observation that Fe absorption responds inversely to the size of Fe stores (Arredondo & Nunez 2005).

The more common problem for humans with iron deficiency is anaemia. The dietary intake level for iron that is sufficient to meet the nutrient requirements of nearly all healthy individuals in a life stage and gender group, which is called recommended dietary allowance (RDA), range from 7 to 27 mg/day depending on age, gender and physiological status (IOM 2001). However, a normal diet generally provides the iron that is needed (Dvorak *et al.* 2007).

Symptoms associated with Fe deficiency are tiredness, lassitude, and general feelings of lack of energy, most of them having their origin in lack of transported oxygen to tissues. In addition, the immune system may be depressed, since iron is essential for proper enzyme function in immune cells and signalling pathways (Bowman & Russell 2006). Fe deficient infants may suffer from delayed motor maturation and cognitive dysfunction, impaired growth and general weakness, and depressed intestinal function (Dallman 1986; Ryan 1997).

Iron toxicity

Iron deficiency is more common than overload. Children and pregnant women usually suffer from anemia, a general feeling of lack of energy. Acute iron poisoning occurs mostly accidentally, when large numbers of iron containing pills are ingested usually by children. It is followed by symptoms like severe vomiting, diarrhea, abdominal pain but also it may cause conjunctivitis, choroiditis, and retinitis if it contacts and remains in the tissues. The average lethal dose of iron is 200–250 mg/kg of body weight, but death has occurred following the ingestion of doses as low as 40 mg/kg of body weight (WHO 2003).

Inhalation of excessive concentrations of iron oxide may enhance the risk of lung cancer development in workers exposed to pulmonary carcinogens.

Chronic Fe intoxication occurs frequently associated to genetic and metabolic diseases, repeated blood transfusions, or excessive intake (Fraga & Oteiza 2002).

It is possible, unfortunately, that a portion of the people may be prone to Iron overload or Hemochromatosis. The symptoms of hemochromatosis can include: chronic fatigue, arthritis, heart disease, cirrhosis, cancer, diabetes, thyroid disease, impotence, and sterility. Hereditary hemochromatosis (HFE) is only one of several iron loading diseases and it is the most common genetic disease but also tragically the most undiagnosed. All ethnic groups can be affected but people of Celtic heritage, with ancestors from Ireland, Wales, Scotland and England, are at high risk because of carrying the HFE mutations for hereditary hemochromatosis (AHS 2012).

An intake of 0.4–1 mg/kg of body weight per day is unlikely to cause adverse effects in healthy persons (WHO 2003). With iron water concentration of 0.3 mg/l, body weight of 60 kg and daily water consumption of 2l, an intake of 0.01 mg Fe/kg of body weight comes from drinking water.

The lack of mechanisms for cell secretion and body excretion causes Fe to accumulate in cells and the body. In cells this gives rise to the labile or reactive Fe pool that is probably involved in the induction of oxidative damage to vital cellular components, whereas body Fe accumulation may give rise, in time, to neurodegenerative and aging processes (Arredondo & Nunez 2005).

Based on a provisional maximum tolerable daily intake (PMTDI) of 0.8 mg Fe/kg of body weight given by the Joint Expert Committee on Food Additives (JECFA) and an allocation of 10% of this PMTDI to drinking-water gives a value of about 2 mg/litre, which does not present a hazard to health (WHO 2004, 2006). WHO has still not set a health-based guideline value for iron, because at the levels found in drinking-water no adverse effects have been found (WHO 2011b).

Recent studies from Sweden (Rosborg 2008) suggest that ingestion of iron in high concentrations from drinking water may be a reason of serious health problems. It was found that people who used to drink water with iron concentrations reaching 5.5 mg Fe/l, suffered from severe constipation, aching neck, hips and arms, head ache and gastric ulcer. In addition muscle cramps during nights and rapid pulse, at instants were observed. It was noticed that in some cases rusty sweet appeared and white shirts became discoloured. Other symptoms such as diarrhoea, lethargy, loss of appetite and weight, apathy, were also observed.

However, it should be emphasised that the levels of iron in drinking water, mentioned in the study were exceptionally high (over 25 times higher than EU parametric value).

KEY POINTS:

(1) Both iron and manganese are essential to human health.
(2) Iron deficiency is more common than overload, however manganese deficiency is rare.
(3) Acute iron poisoning occurs mostly accidentally. There is no evidence for iron toxicity from drinking water at the concentrations commonly found in potable water.
(4) The toxic effects of iron and manganese depend on its exposure level, route of exposure, chemical form, the age, nutritional, health status and inter-individual variation.
(5) Compared to adults at the same manganese drinking water concentration, exposure to Mn is higher in case of infants and children, due to the greater water consumption relative to their body weight.
(6) Infants and young children are more sensitive to adverse health effects of manganese than adults, so it should be taken into account in health risk assessment.
(7) Although many epidemiological studies have reported adverse neurogical effects of Mn in drinking water at various concentrations, however, there are many uncertainties concerning the health effects of Mn and its possible dose-response relationships.
(8) Presence of iron and manganese in drinking water at levels well below those of heath concern would affect consumer acceptance of water.

Chapter 6

Aesthetic issues

6.1 ACCEPTABILITY OF DRINKING WATER

Drinking-water should not only be safe but also acceptable in appearance, taste and odour. This provision is given a high priority by the World Health Organization (WHO 2011a).

Water that is aesthetically unacceptable will undermine the confidence of consumers and will be a reason for complaints or choosing some sort of alternative to tap water, for example, bottled water (Levallois *et al.* 1999). Average consumers attitude towards their drinking water supply depend on aspects that they are able to perceive with their own senses.

On the other hand acceptable aesthetic quality does not necessarily make water safe. Water that is aesthetically acceptable might contain organoleptically undetectable toxic elements and/or be microbiologically contaminated.

The most important acceptability aspects are:

- Colour.
- Taste.
- Odour.

Discolouration along with "taste and odour" incidents is among the most frequent reasons for consumers' complains worldwide (Krasner *et al.* 1983; Ventura *et al.* 1995; Pattison *et al.* 1995; Sarin *et al.* 2001, 2004; McNeill & Edwards 2001; Vreeburg *et al.* 2008; Doria *et al.* 2009; Husband & Boxall 2008, 2011).

Usually consumers regard with suspicion water that appears dirty or discoloured or that has an unpleasant taste or smell especially if changes in water appearance occur suddenly. These characteristics may not in themselves be of direct consequence to health, but the water may be rejected, as unacceptable for drinking purposes. The acceptance level is variable and dependent on individual and local factors, including the aesthetic quality of the water to which the community is accustomed. Other factors including social, environmental and cultural ones play also an important role.

6.2 COLOUR

Drinking water should have no visible colour. Colour in drinking water is usually due to the presence of coloured organic matter associated with the humus fraction of soil. Colour may be strongly influenced by the presence of iron and manganese in water, either of natural origin or as corrosion products.

Groundwater may contain ferrous iron at concentrations up to several milligrams per litre without discoloration or turbidity in the water when directly pumped from a well. On exposure to atmospheric oxygen the ferrous iron oxidizes to ferric iron, giving a reddish-brown colour to the water. Iron also promotes the growth of "iron bacteria", which form a slimy coating on the piping. At levels above 0.3 mg/l, iron stains laundry and plumbing fixtures.

The presence of manganese in drinking water may lead to the accumulation of deposits in the distribution system. At a concentration of 0.2 mg/l, manganese may form a coating on pipes, which may slough off as a black precipitate giving black colour to the water. Manganese can also stain sanitary-ware, shower trays and baths.

The source of colour in a drinking-water supply should be investigated, particularly if a substantial change has taken place.

Corrosion of cast iron is often seen as the main source for discolouration problems. However, the particles originating from the treatment plant play an important and potentially dominant role in the generation of a discolouration risk in drinking water distribution systems (Lytle *et al.* 2004, 2005; Vreeburg *et al.* 2008; Vreeburg 2007, 2010; Husband & Boxall 2011). In the Netherlands, less than 10% of the network consists of unlined cast iron and the water is biological stable (van Dijk & van der Kooij 2005). Every year 3000–6000 complaints about discoloured water reach the water companies. That means 1 complaint per 1000 connections per year (Vreeburg *et al.* 2008). The colour of a water sample can be reported as:

- *Apparent colour* is the colour of the whole water sample, and consists of colour from both dissolved and suspended components.
- *True colour* is measured after filtering the water sample to remove all suspended material.

Most people can detect colour above 15 true colour units (TCU) in a glass of water. Levels of colour below 15 TCU are believed to be acceptable to consumers (WHO 2011a). An average consumer can only asses the apparent colour (see Figure 6.1).

Figure 6.1 Difference between apparent colour and true colour of the same water.

Red water phenomenon

Iron release from corroded iron pipes is the principal cause of "red water" problems in drinking water distribution systems (Baylis 1926; Larson 1939; Larson & Skold 1957; McNeill & Edwards 2000, 2001; Clement *et al.* 2002; Imran *et al.* 2005a; Lytle *et al.* 2005; Zhang & Edwards 2007).

Iron release occurs when there is a transport of iron from corrosion scale or the metal surface to the bulk water, either in soluble or particulate form. Coloured water is formed when iron enters the bulk water as ferric particles, or as Fe(II) that oxidizes and then forms ferric particles.

In the absence of any corrosion scales, corrosion of iron is the primary cause of iron release. When metal surfaces are covered with corrosion scales, iron may be released by:

- The corrosion of iron metal.
- The dissolution of ferrous components of the scales.
- Hydraulic scouring of particles from the scales.

Interactions of corroded iron surfaces with water quality are not completely understood. In corroded pipes there is no simple relationship between iron release and the rate of iron corrosion. It is recognized that water quality parameters can influence iron release from corroded iron and steel pipes in distribution systems. The most important factors influencing iron release are:

- Alkalinity.
- Application of an inhibitor and type of inhibitor used.
- Buffer intensity.
- Dissolved oxygen (DO).
- Fluctuations in water quality (chlorine, chlorides, sulphates).
- Bacterial activity.
- pH.
- Temperature.
- Water flow characteristics.
- Water treatment practices.

Imran (*et al.* 2005b) developed a colour release model based on influent water quality using weekly average data from 14 pilot distribution systems. This model showed that apparent colour could be closely predicted based on the set of water quality parameters and the change in apparent colour: effluent – influent can be calculated from the formula:

$$\Delta C = \frac{(Cl^{-})^{0.485} \times (Na^{+})^{0.561} \times (SO_4{}^{2-})^{0.118} \times (DO)^{0.967} \times (HRT)^{0.836}}{20.7 \times (ALK)^{0.912}}$$

where:

ΔC is the change in apparent colour: effluent – influent (CPU)
Cl^{-} is the chloride concentration in mg/l
Na^{+} is the sodium concentration in mg/l

SO_4^{2-} is the sulphate concentration in mg/l
DO is the dissolved oxygen concentration in mg/l
T is the temperature in °C
HRT is the hydraulic retention time in days
ALK is the alkalinity as mg/l $CaCO_3$

This model showed relatively good correlation ($R^2 = 0.83$) with colour data, and was verified using an independent data set. The data also showed that waters low in alkalinity (<80 mg/l as $CaCO_3$) corresponded to significantly higher concentrations of effluent iron. Water quality parameters such as chlorides, sulphates, and dissolved oxygen were statistically significant in elevating effluent colour levels based on the model.

Existing theories however, are not sufficient to fully explain "red water" and excessive iron corrosion problems. Corrosion control measures are still often based on trial and error approaches.

Black water phenomenon

Manganese present in water, even at concentrations as low as 0.02 mg/l, can form a coating of manganese oxides on pipes (Griffin 1960). It could be a result of inadequate removal of manganese at the water treatment plant. Soluble manganese enters the distribution system where it is oxidized to its insoluble form either by residual concentrations of chlorine or by bacteria, depositing these oxidised solids on the pipe surfaces (Sly *et al.* 1990). This deposit may be released to bulk water as a result of fluctuations in water flow regime and cause black discolouration.

The formation of manganese-reach coatings could take place in all types of pipes that are commonly used in distribution system including cast iron, iron, PVC and HDPE (Cerrato *et al.* 2006; Murdoch & Smith 1999). However, it was documented (Cerrato *et al.* 2006) that in iron pipes manganese was incorporated into the iron tubercles and thus not readily dislodged from the pipes by water flow. In PVC pipes the manganese-rich scale was easily dislodged by flowing water.

Increasing chlorination levels during treatment, to control manganese-depositing biofilm in water distribution pipes, will not always reduce manganese deposition. When excessive Mn(II) continues to enter the distribution system increased chlorination levels may considerably worsen chemical oxidation and deposition. As a result it may cause more serious and widespread dirty water problem because the rate of chemical deposition is considerably greater than microbial deposition (Sly *et al.* 1990).

Deposits of manganese in distribution systems are generally considered to be harder to remove than iron deposits, due to the fine, dense nature of the manganese coating. Often, flushing of mains will not be sufficient and air scouring, swabbing or even rehabilitation techniques may be required.

6.3 TASTE

The taste of water may be detectable by consumers at lower or higher levels, depending on individual and local circumstances. It is one of the most "subjective" parameters of water quality. However, it may be indicative of some form of pollution or of problems during water treatment or distribution.

There is usually no noticeable taste at iron concentrations below 0.3 mg/l (WHO 2011a). At higher concentrations a metallic taste may appear. A bittersweet astringent taste is detectable at concentrations above 0.5 mg/l by many people (Juyun & Lawless 2005; Epke & Lawless 2007). When water with high iron content is used for tea or coffee and other beverages preparation, it may produce a harsh, unacceptable taste (Colter & Mahler 2006)

At levels exceeding 0.1 mg/l, manganese in water supplies causes an undesirable taste in beverages (Griffin 1960). Concentrations below 0.1 mg/l are usually acceptable to consumers (Casey 2009). The health based value of 0.4 mg/l set for manganese by WHO (2011b) is higher than the acceptability level (see Chapter 5 for more information).

6.4 ODOUR

Odour in drinking water may be an indication of some form of pollution of the water source or of a malfunction during treatment or distribution. It is not possible to specify a maximum acceptable limit for odour, because no method for its objective measurement exists (Health Canada, 1995; Rogers 2002).

Odour in water is usually assessed in terms of its threshold odour number (TON) – the number of times a sample must be diluted with an equal volume of odour-free water to become just detectable. An alternative measurement, which is known as the odour intensity index (OII), is used on some occasion in case of substances that can cause very intense odours at exceptionally small concentrations. The OII represents the number of times the volume of a sample must be diluted with odour-free water before it is just detectable. Flavour profile analysis (FPA) is another technique for identifying sample odour(s). This procedure is described as standard method 2170 (SMEWW 2170). FPA differs from threshold odour number. The sample is not diluted and each taste or odour attribute is individually characterized and assigned its own intensity rating.

Odour assessment using sensory methods is a basic and simple test however, it is essentially a subjective assessment therefore has considerable uncertainty. Published standard methods for the assessment of tastes and odours in drinking water suffer from methodological and interpretational problems (Peryam 1958; Lin 1977; SMEWW 2150; Young *et al.* 1996; Bruchet *et al.* 1995).

Iron and manganese are not direct sources of any odour, but since manganese can be removed from water by chlorination followed by filtration, occasional problems caused by chlorination residuals may appear. Odours caused by disinfectants are best controlled through careful operation of the disinfection process and pre-treatment.

Iron bacteria are widely distributed in nature, and commonly observed in water distribution systems (see Chapter 3 for more information). Iron bacteria derive their energy from the oxidation of iron and manganese. Products of these reactions are accumulated on the sheaths and cells of the organisms. Dying bacteria leave a reddish brown or yellow slime that can clog plumbing and cause an offensive odour (Lueschow & Mackenthun 1962; Shalini & Pragnesh 2012). Common iron bacteria odours include: swampy, rotten egg, sewage, fuel oil and cucumber. These odours can sometimes be confused for different contaminants like hydrogen sulphide gas. The odour caused iron bacteria activity may be worse if water hasn't been used for a prolonged period of time.

Another source of strong odour, associated with iron and manganese presence, is hydrogen sulphide which commonly occurs in ferruginous groundwaters (usually extracted from deep aquifers). The rotten-egg odour of hydrogen sulphide can be detected by most people in waters that have only a few tenths of a milligram per litre of this material in solution (Hem 1985). It results from anaerobic bacterial action on sulphates. Reduction processes that can occur where oxygen is depleted may convert sulphate to sulphur or sulphide. The sulphide may then react with ferric iron and reduce it to ferrous iron (Hem 1985; Gault *et al.* 2009).

KEY POINTS:

(1) Discolouration of water and other aesthetic aspect are the most common reason for consumers' complaints.
(2) The colour in drinking water is usually due to the presence of coloured organic matter associated with the humus fraction of soil but it may be strongly influenced by the presence of iron and manganese in the water, either of natural origin or as corrosion products.
(3) In most cases there is no direct relation between water quality and its appearance. However, water with high Fe or/and high Mn contents will show visible discolouration.
(4) Proper assessment of the safety of drinking water, just on the base of aesthetic quality is not possible. Satisfying aesthetic quality does not necessarily make water safe. Water that is aesthetically satisfactory might be microbiologically contaminated or contain undetectable with human senses toxic compounds.

Chapter 7

Sampling and monitoring

7.1 ROLE OF SAMPLING AND MONITORING

Typical sampling program objectives are:

- To assess compliance with drinking water quality standards.
- To determine the suitability of raw water for treatment to the appropriate standards.
- To support treatment process design.
- To assess the performance and control of treatment plants.
- To assess water quality changes in distribution systems.

The objectives rarely stand alone and will affect each other, for instance if legal quality standards are applied at household taps then treatment objectives will have to allow for any effects of the distribution system. For example, iron concentrations will often rise during distribution due to the corrosion of iron and steel pipes (Larson 1939; Larson & Skold 1957; AWWARF 1996; McNeill & Edwards 2001; Sarin *et al.* 2004a, b; IWA 2012a).

The most obvious consequence of poor sampling and monitoring is that we may unknowingly be breaching one or more regulatory standards. This can be caused by inadequate sampling frequency (i.e. we miss the peak values), inadequate sampling procedures or poor choice of monitoring points (IWA 2012c). Inadequate sampling and monitoring can lead to unacceptable aesthetic quality of the drinking water and/or hazards to public health. The other important point is that water companies may make unjustified decisions regarding changes to the waterworks treatment which may result in unnecessary expenditure.

Regardless of the purpose of monitoring adequate procedures of sampling program design and sample collection protocols should be followed. Sampling and monitoring procedures are usually formulated by health and environmental authorities but majority of them are based upon ISO standards. Most important ISO standards regarding sampling and monitoring in drinking water supply systems are assembled in International Standard ISO 5667 – Water Quality – Sampling, which are:

- Part 1: Guidance on the design of sampling programs and sampling techniques.
- Part 3: Guidance on the preservation and handling of water samples.

- Part 4: Guidance on sampling from lakes, natural and man-made.
- Part 5: Guidance on sampling of drinking water from treatment works and piped distribution systems.
- Part 6: Guidance on sampling of rivers and streams.
- Part 11: Guidance on sampling of groundwaters.
- Part 14: Guidance on quality assurance of environmental water sampling and handling.
- Part 20: Guidance on the use of sampling data for decision making – Compliance with thresholds and classification systems.
- Part 22: Guidance on the design and installation of groundwater monitoring points.
- Part 23: Guidance on passive sampling in surface waters.

7.2 DESIGN OF SAMPLING PROGRAMMES

Sampling and monitoring programs should be designed, detailing the strategies and procedures to follow, for monitoring the various aspects of the water quality within the supply system from source to the tap (see IWA 2012c). Programs should be designed to cover both spatial and temporal variations in water quality, to give information representative of the quality of the water supplied to consumers. The monitoring programs should be designed by personnel experienced in the assessment of water quality, and should be fully documented.

Prior knowledge of the quality of a water body may be used to help design a sampling program for a new objective. However, if we have insufficient or no prior knowledge of quality then it becomes very difficult to design an effective sampling program. In such cases a preliminary sampling program should be undertaken in sufficient detail to design a program which will enable the defined objectives to be met at least cost (IWA 2012c, ISO 5667).

In order to obtain comparable results it is often necessary to sample from a given location at specified intervals. It is therefore essential to be able to locate the sampling point appropriately.

Iron and manganese concentrations may vary widely, rapidly and continuously or intermittently and be affected by such factors as time, flow rate and plant operation. In some cases it will be necessary to sample continuously to achieve an objective, however, this is generally expensive and often not necessary and most objectives can be achieved by appropriate programs of discreet sampling. Continuous monitoring of iron at a waterworks can be linked to coagulation process control. There will be a need to allow for seasonal cycles where factors such as temperature and rainfall could affect concentrations, diurnal (daily) cycles such as variable pumping from a borehole and weekly cycles (business week) which will affect the pattern of household water usage. In addition there may be random or transient events of interest and long-term trends (Macioszczyk 1987; Nielsen 2006).

7.3 RAW WATER SAMPLING

Raw water is the natural resource that has not been purified, such as water within a lake, pond, river, reservoir or an aquifer. From a drinking water production point of view raw water is commonly understood as the water entering treatment works. Samples would normally be taken as the water enters the waterworks; however provision should also be made to enable sampling upstream, such as at the wellhead or the river above the abstraction point. An accurate knowledge of the chemical composition of water entering a treatment plant is essential for the design of treatment processes and for operational purposes.

The primary objective of most raw water sampling programs is to collect samples that are representative of the water in its natural condition. Different sampling methods may need to be used in order to collect representative samples that will properly reflect the chemical composition of a relatively large body of water. The methods used to collect water samples depend on flow characteristics of the water body, temporal and spatial heterogeneity; physical setting; and weather conditions.

Sampling and monitoring of raw waters used for the production of drinking water is generally driven more by environmental considerations than drinking water treatment needs (Skotak *et al.* 2007; Nielsen 2006). The main objective is to protect the natural water resources that are used for drinking water supplies. If the monitoring is properly performed it should help to predict potential hazards for water quality and to steer preventive action.

Surface waters

Surface water quality may vary significantly in space and time. It is necessary to take into account the influence of field conditions on the chemical analyses of natural water in terms of spatial and temporal variability. Surface waters are also, from their nature, vulnerable to pollution incidents, which may rapidly change the chemical composition of the water. Surface water catchments may be represented by lakes, ponds, reservoirs and rivers and streams. Each type of catchment will present different challenges to proper sampling.

In most cases, one sample of the water will not be representative of the entire body. Samples should be taken at various locations in the horizontal and vertical direction (usually from a boat). Samples of the water being pumped to the waterworks are taken at the works before any chemicals are added. Samples may also be taken at the intake or at appropriate points upstream of a river abstraction. In the case of lakes and ponds there is often only one point from which the water is pumped to the waterworks. In reservoirs there is usually an intake tower located close to the deepest part of the reservoir and water can be pumped from various depths to obtain the best quality of water for treatment.

The discharge of a river is related to the nature of its catchment, particularly the geological, geographical and climatic influences. Different aspects of river flow such as annual water yield and its temporal variations, drogue periods and storm flow events should be considered. Changes in water quality may occur at different rates and change to different degrees. The chemical composition of river water may vary due to seasonal, weekly and/or diurnal (daily) cycles and may change rapidly during pollution incident (Hem 1985; Neal *et al.* 2005; Alvarez-Vazquez *et al.* 2006; Naddeo *et al.* 2007). An important concept for water pollution studies is residence time. It is associated with the time taken for recovery from a pollution incident. Rivers are characterized by a short residence time. That aids recovery of the aquatic system from a pollution input by rapid dispersion and transport of waterborne pollutants, but requires more frequent monitoring. Knowledge of the time of travel of a pollutant along a river will help the design of contingency measures. In some cases continuous monitors may be useful with alarms linked by telemetry to a central control facility.

Collecting spot water samples of surface water is relatively simple. In most cases it can be done using a bucket or wide-mouthed bottle dropped or immersed into a body of water. If it is not possible, because of local conditions or special requirements, samplers of different design can be used. The sampler is usually an open tube made of inert material, with spring-loaded end caps. There are several types of water samplers that can be used for collecting water samples at discrete depths within a body of water which is especially important when sampling stratified bodies of surface water (IWA 2012c). Examples of typical samples are presented on Figure 7.1.

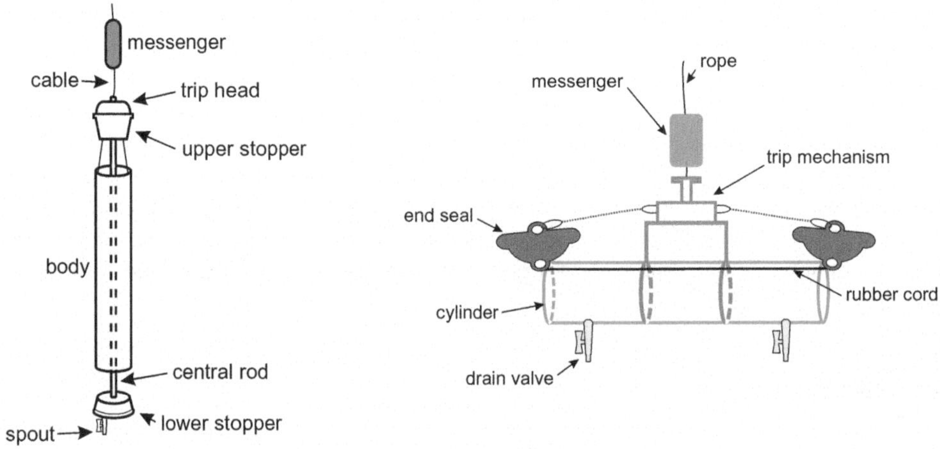

Figure 7.1 Typical samplers used in surface water sampling (after IWA 2012c).

Groundwaters

Samples are normally obtained from existing, working wells. The usual situation is that a producing well will be a groundwater quality monitoring point. Where the wells are new or have been stagnant for substantial period of time, samples should be collected after pumping a significant volume of water from the well. This is called purging.

When a particular sampling point is not equipped with a pump portable pumps or samplers or bailer may be used for purging and sample collection. Bailers are commonly used for both purging and sampling in small diameter, shallow wells, primarily because of their convenience, ease of use, and low cost. However, when the bailer is used for purging and then sample collection, significant problems with increased values of dissolved oxygen and turbidity may occur (Stolzenburg & Nichols 1986; Pohlman *et al.* 1994).

Groundwater samples can be collected using different types of portable pumps, such as peristaltic pumps or bladder pumps. The pump material should as far as possible, be inert and not give raise to any interaction between the water and the material of construction. It is also very important that devices used for purging and sampling have a controllable flow rate. Peristaltic pumps have the advantages of few moving parts, easily replaceable heads, and portability. Bladder pumps provide representative samples and are recommended for sampling in variety of field conditions (e.g. Barcelona *et al.* 1983; Hudak 2005; Parker 1994; Pohlmann *et al.* 1994). Other types of pumps may also be used for purging and sampling both as dedicated or portable devices. Detailed descriptions of many available types of pumps, with their advantages and disadvantages, can be found in numerous publications (e.g. Nielsen 2006; Parker 1994).

To obtain representative samples and prevent sample alteration, subsurface disturbance and sample handling must be kept to a minimum. Field measurements of physical and chemical parameters (temperature, pH, Conductivity, Eh, dissolved oxygen) provide important information on water quality and can be used as indicator parameters for defining stabilization of water-quality during purging. Water temperature, especially the temperature of shallow ground water, is approximately equal to the mean annual air temperature. It may serve as an indicator of stability in shallow ground-water chemistry. All measurements should be performed under conditions that avoid any contact between water and atmospheric oxygen (Nielsen 2006). For anaerobic waters it may be necessary to use sampling techniques which prevent contact between the sample and the atmosphere in order to avoid sample alteration specifically in regard to iron.

Sampling frequency

The frequency with which raw water samples are collected and used for iron and manganese content analysis is usually determined by the regulatory agency (e.g. Sanitary Inspection), the water quality historical record, and treatment works operational control requirements.

Operators of public water systems and industrial and commercial water systems will want to collect more frequent but carefully selected samples and make more analyses to detect changes in raw water quality to better control treatment, plant operation, and product quality. Resource constraints often apply to small private or community water supplies. The sampling frequency should account for:

- Seasonal cycles.
- Diurnal (daily) cycles.
- Weekly cycles (business week).
- Random or transient events.
- Long term trends.

Under abnormal conditions increased sampling frequencies may be used. If these results are included in the calculation of long term averages and trends, allowance must be made for the increased frequency by time weighting.

If raw water quality changes rapidly (e.g. river or stream water), continuous monitoring should be considered.

7.4 SAMPLING AT WATER TREATMENT WORKS

Iron may occur naturally in raw water or may be added as part of the treatment process. In order to properly monitor and control the treatment process it is important to measure the concentration of iron and other relevant metals in the water at various stages through the process. The point from which a sample is taken is crucial in determining whether or not a sample is representative of the main body of water.

Sampling of iron and manganese during and after the water treatment process takes place for a number of reasons:

- Performance monitoring.
- Process control.
- Regulatory compliance.
- Investigatory.

Care should be taken when sampling from pumped mains on water treatment works. Such mains are best avoided if possible, as water in them may be standing for a period of time whilst pumps are not operating, leading to unrepresentative sampling. Similarly, turbulence caused by pumps starting up may disturb sediment in the main that can contaminate the sample.

Sample lines should be constructed from an inert material that will not react with the water or chemicals added to it and should be connected to the main body of water. Any areas of slow, laminar flow within the pipe should be avoided as these could lead to the deposition of sediments in the pipeline.

On-line monitors can be used to automatically sample and analyse a water stream for selected parameters, including iron and manganese concentrations. Monitors can provide an almost continuous data stream that enables transient problems to be identified, such as metal breakthrough occurring during a filter wash.

7.5 SAMPLING METALS IN DISTRIBUTION SYSTEMS

The term distribution system refers to the water supply network between the treatment works to the point at which the water enters the user's pipe-work. It includes large diameter trunk mains, smaller distribution mains and any storage points such as service reservoirs and water towers.

Iron and manganese may enter the system from the treatment works, either due to inadequate treatment or poor performance of a treatment process. Iron and manganese may also derive from corrosion of the water mains material itself, and the iron content in the water may increase significantly as it passes through the distribution system. Some deposits of manganese in the distribution system may be due to historic issues with treatment or water mains, even if the original source is no longer present.

As with all sampling, care needs to be taken to ensure that the sample obtained is representative of the body of water. This is especially true in distribution systems, where concentrations of metals can vary across short distances, and the opportunities to obtain a sample may be limited.

Sampling in the distribution system is often used to provide an indication of the quality of water across a whole area. For many metals, especially iron and manganese, the geographical variation can be quite considerable as metal concentrations can depend on mains material and sediment, which can accumulate at certain locations within a system. Areas of high concentration can travel through the system as a wave due to pressure fluctuations.

All these factors need to be taken into consideration when designing and undertaking a sample survey, to ensure that the results obtained are suitable for the intended purpose. In particular, the size of the survey is important to ensure adequate coverage of the target area and to enable meaningful statistical analysis to be carried out on the result.

Reasons for sampling and monitoring in distribution system include (IWA 2012c):

- Investigatory – undertaken in order to identify and resolve a specific issue for example, regulatory failure or consumer's complaint.
- Incident management – undertaken in order to determine extent and scale of a water quality incident at the treatment works.
- Regulatory compliance – sampling undertaken to comply with national or regional regulations.
- Operational – to determine and optimize the performance of the distribution system.
- Investment planning and appraisal.

Sampling locations

The easiest method of obtaining a sample directly from the distribution system is using hydrants and standpipes, although sample locations are limited by the location of the correct mains fitting.

When sampling from a hydrant it is important to be aware of the impact of the sampling activity itself on the quality of the sample obtained. Use of a small diameter standpipe and gradual operation of valves to obtain a steady flow of water is important in order to avoid disturbing sediment within the main. The sample, collected under steady state conditions is most representative of water normally supplied to consumers.

In some cases it may be desirable to obtain samples under "non-steady-state" conditions in order to establish a worst case scenario of quality in the system during times of disturbed flows. Such sampling however, needs to be managed very carefully to minimize the effect on consumers (IWA 2012a, c).

Many modern designs of meter chamber or boundary box provide the opportunity for a sample to be taken since many of them are equipped with a sampling tap. These sample locations have the benefit of being relatively common and provide the opportunity to obtain a sample at the last point of the distribution system before water enters the consumer's pipe-work. Such water samples may provide important information on chemical processes occurring in distribution network and within the consumer's pipe-work when confronted with the results of analyses of samples collected at consumers' taps and water entering the distribution system. If the collection of samples is not possible at a meter chamber or boundary box, samples can be taken at a consumer's tap after prolonged flushing. This sample will also reflect the quality of water in distribution system.

7.6 SAMPLING AND MONITORING AT CONSUMERS' TAPS

The consumer's tap is the monitoring point of greatest relevance to drinking water consumption. It is usually considered as the point of compliance for regulatory monitoring. This is normally taken to be the cold water tap in the kitchen where water is used for drinking and food preparation.

Iron and manganese concentrations in drinking water at the consumer's tap tend to vary substantially, for two major reasons:

- Metals that arise at the source and within the distribution network have a tendency to settle, accumulate and then be re-suspended, depending on water flow characteristics.
- Metals that arise from domestic pipe-work systems are dependent on the highly variable contact time between the water and the pipes or fittings.

Sampling protocols

Different sampling protocols can be used when sampling at consumers' taps depending on the purpose of investigation. The most commonly used types of samples (sampling protocols) are:

- Fully flushed samples (FF) – samples are taken after prolonged flushing of the tap in a premise in such a way that stagnation of water in the domestic distribution system does not influence the concentration of metals. The main value of a fully flushed sample is the ability to confirm the extent of problems with metals derived from source or the distribution network. The chemical composition of such a sample may not be representative of the water being consumed.
- First draw samples (FD) – samples taken first in the morning before the tap in the premise has been used for other purposes.
- Composite proportional samples (COMP) – samples are collected during a defined (e.g. 1-week) period. This sampling method provides a direct measure of the average metal concentration being emitted at the consumer's tap and will approximate to the amount of metals ingested. This sampling method requires a special consumer-operated device to be fitted to the tap that splits off a small constant proportion of every volume of water drawn for dietetic purposes (Figure 7.2).
- Stagnation samples – samples taken after prolonged flushing of the tap in a property and successive stagnation for a predefined period before the sample is taken from the tap. A stagnation time of 30 minutes and a 1 litre sample volume are commonly used, but stagnation time varies in different countries from 30 minutes (30MS) to 12 hours (IWA 2012c).
- Random daytime samples (RDT) – samples taken at a random time of a working day directly from the tap in a property without previous flushing. It is common practice to select the sampling points at random and for 1 litre sample volumes to be taken. When sampling locations are selected at random, the method provides an unbiased assessment of zonal compliance. Random daytime sampling does not suffer from the potential problems with selecting representative sampling locations. It is also the most logistically favourable method as the time involved in sampling is

very short and it is acceptable for the sampler to optimize the sampling route on each day to optimize travelling. The main weakness of random daytime sampling is that sample numbers need to be sufficiently high to achieve confidence in the assessments made. RDT samples can have poor reproducibility (van den Hoven *et al.* 1999) if sample numbers are low. However, the UK has successfully used RDT sampling for metals at the tap for regulatory purposes since at least 1989 (some regions adopted RDT sampling from 1984).

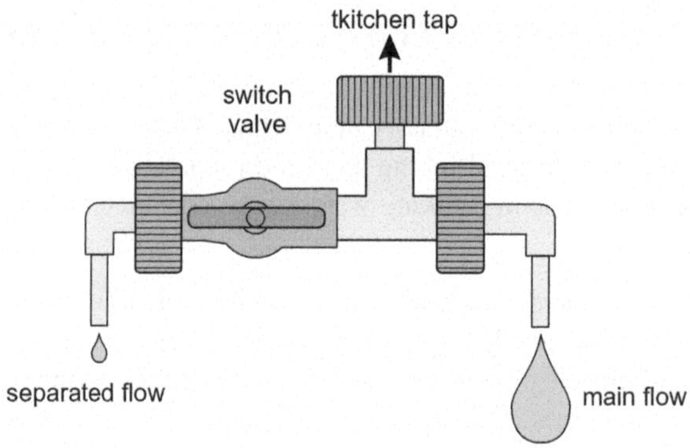

Figure 7.2 Diagram of split-flow sampling device.

Sampling frequencies in compliance monitoring

The regulations that establish standards for metals in drinking water require compliance to be demonstrated by monitoring. Compliance monitoring is the monitoring at the point of compliance to verify that water supplied for human consumption is in compliance with its quality requirements. Regulatory compliance monitoring tends to focus on the points across an entire water supply system (98/83/EC).

In the EU, a minimum number of samples must be taken each year from consumers' taps for a either check or audit monitoring. The purpose of check monitoring is to provide information on the organoleptic and microbiological quality of the water supplied as well as information on the effectiveness of drinking water treatment. The purpose of audit monitoring is to provide the information necessary to determine whether or not all the Directive's parametric values are being complied with. All parameters set in accordance with Article 5(2) and (3) must be subject to audit monitoring (unless it can be established that a parameter is not likely to be present in a given supply in concentrations exceeding parametric value). Required frequencies for check and audit monitoring are given in Table 7.1.

Minimum frequencies for compliance monitoring should depend on the risk assessment made in the water risk management strategy and could be higher. As far as possible, the number of samples should be distributed equally in time and location.

Member State may reduce the number of samples specified in the table if the results obtained from samples taken during a period of at least two successive years are constant and significantly better than the parametric values and there are no factors likely to cause a deterioration of the water quality. However, the lowest frequency applied must not be less than 50% of the number of samples specified in the table.

It is recognized that these minimum frequencies (Table 7.1) are inadequate for operational control purposes (Hoekstra *et al.* 2008).

Table 7.1 Minimum monitoring frequencies for iron and manganese in the EU.

<table>
<tr><th>Volume of water distributed</th><th>Check monitoring
Fe[1]</th><th>Audit monitoring
Fe, Mn</th></tr>
<tr><td>[m³/day]</td><td>[samples/year]</td><td>[samples/year]</td></tr>
<tr><td>≤100</td><td>The frequency is to be decided by the Member State</td><td>The frequency is to be decided by the Member State</td></tr>
<tr><td>>100 – ≤1000</td><td>4</td><td>1</td></tr>
<tr><td>>1000 – ≤10,000</td><td rowspan="3">4
+ 3 for each 1 000 m/d and part thereof of the total volume</td><td>1
+ 1 for each 3,300 m³/day and part thereof of the total volume</td></tr>
<tr><td>>10,000 – ≤100,000</td><td>3
+ 1 for each 10,000 m³/day and part thereof of the total volume</td></tr>
<tr><td>>100,000</td><td>10
+ 1 for each 25,000 m³/day and part thereof of the total volume</td></tr>
</table>

[1] – Necessary only when Fe is used as flocculants.

Special investigations at individual properties

Investigations at individual properties should be undertaken following consumers' complaints or if the results of regulatory monitoring fall outside pre-determined limits. Non-compliance with the relevant regulatory standard at consumers' taps may reveal atypical circumstances such as recent plumbing alterations or significant changes in water flow regime. Judgment and pre-agreed protocols will be needed to determine if such atypical results should be included in compliance assessment. In the UK, all results from compliance assessment are reported, regardless of circumstances.

An investigation should comprise:

(1) An inspection of the pipe-work from the water main to the kitchen tap to record pipe materials, lengths and diameters and age of pipe-work.
(2) Enquiries to determine if any recent pipe-work changes have occurred.
(3) Observation of proximity to heavy road traffic, construction works and possible vibration.
(4) Assessment of general water quality characteristics in the area.
(5) Checking the occurrence of water discolouration episodes.

Taking repeat random daytime samples, to attempt to confirm the first sample result, is not recommended, because metal emissions from taps vary too much. A combination of samples collected according to different protocols may provide a better measure of the extent of metal emission problems and will help to locate the source of a particular problem (see Table 7.2).

When investigating iron and manganese problems additional water quality parameters should be analysed, which are: water temperature and pH (*in situ*), alkalinity, calcium, chlorides and sulphides and dissolved oxygen (DO). All these parameters may have strong influence on corrosion of iron pipes which leads to deterioration of water quality especially in regard to aesthetic quality (Larson & Skold 1957, 1958; US EPA 1982, 1984; Dodrill & Edwards 1995; Volk *et al.* 2000; McNeil & Edwards 2001, 2002; Sarin *et al.* 1995, 2004b).

Table 7.2 Sampling framework at individual properties.

Sampling protocol	Information provided	Disadvantages
COMP	Average metal concentrations	Logistics and quality assurance constraints
RDT	Average and range of metal concentrations over a water supply area	Poor reproducibility if sample numbers are low
FD	Maximum concentration of metals arising from domestic pipe-work	Does not determine the load ingested by consumers
FF	Water quality in distribution network	Gives no information on the influence of internal pipe-work

KEY POINTS:

(1) Sampling and monitoring programs should be designed to cover both spatial and temporal variations in water quality, and to give information representative of the quality of water supplied to consumers.
(2) Compliance monitoring is the monitoring at the point of compliance to verify that water supplied for human consumption is in compliance with its quality requirements. Monitoring at consumer's tap is the most important type of water quality monitoring both from the regulatory and human health points of view.
(3) Proportional sampling methods provides a direct measure of the average metal concentration being emitted at the consumer's tap and will approximate to the amount of metals ingested, however, it is time consuming and requires extensive consumer cooperation.
(4) Random daytime sampling provides an unbiased assessment of zonal compliance but a sufficient number of samples must be collected.
(5) Stagnation sampling will provide a better measure of the extent of iron and manganese emission problems from domestic pipe-work.

Chapter 8

Analytical methods

8.1 METHOD SELECTION

Several analytical techniques are available for analysis of iron and manganese content in water. Selecting the most appropriate method is essential for achieving accurate and reliable results. Proper selection requires a basic understanding of each technique since each one has its individual advantages and limitations.

Accordingly to the measurement principle, analytical methods can be divided into following classes:

(1) Colorimetric methods.
(2) Electrochemistry methods.
(3) Atomic absorption spectroscopy.
(4) Inductively coupled plasma spectrometry.

Several questions should be answered when a method for analysis is selected. Usually the first step is establishing what is the purpose of the analyses (legal/scientific/process control)? To answer this question implies consecutive questions to be answered such as:

- Required limits of detection.
- Data quality (trueness, precision).
- Additional legal and/or researcher's requirements.
- Costs involved.

The EU drinking water directive (98/83/EC) requires that: "… any laboratory at which samples are analysed has a system of analytical quality control that is subject from time to time to checking by a person who is not under the control of the laboratory and who is approved by the competent authority for that purpose". It also requires in Annex II 2.1 that the method of analysis used "must be capable of measuring concentrations equal to the parametric value with a trueness, precision and limit of detection specified"(see Table 8.1).

In the United States of America criteria for analysing iron and manganese are contained in Section IV of Technical Notes on Drinking Water Methods – Analytical Methods Recommended for Drinking Water Monitoring of Secondary Contaminants (US EPA 1994) published by US Environmental Protection Agency. Any of these methods can be legally used in drinking water monitoring. Table 8.2 presents US EPA approved methods for iron and manganese.

Table 8.1 Performance characteristics required by directive (98/83/EC).

	Trueness % of parametric value (in μg/l)	Precision % of parametric value (in μg/l)	Limit of detection % of parametric value (in μg/l)
Iron	10 (20)	10 (20)	10 (20)
Manganese	10 (5)	10 (5)	10 (5)

Table 8.2 Analytical methods recommended by US EPA for iron and manganese monitoring in drinking water.

	Method	Method number
Iron (total)	Atomic Absorption direct aspiration	3111 B
	Atomic Absorption furnace	3113 B
	Stabilized Temperature Graphite Furnace Atomic Absorption	200.9
	Inductively Coupled Plasma – Atomic Emission Spectrometry	3120 B, 200.5, 200.7
	Inductively Coupled Plasma – Mass Spectrometry	
Manganese (total)	Atomic Absorption direct aspiration	3111 B, 243.1
	Atomic Absorption furnace	3113 B, 243.2
	Stabilized Temperature Graphite Furnace Atomic Absorption	200.9
	Inductively Coupled Plasma – Atomic Emission Spectrometry	3120 B, 200.5, 200.7
	Inductively Coupled Plasma – Mass Spectrometry	200.8

8.2 COLORIMETRIC METHODS

The principle of colorimetric methods is a reaction of a metal ion with a reagent in order to produce a coloured substance that will absorb light at a specific wavelength, in the UV-Visible region. The intensity of the developed colour is related to the concentration of the metal ion in the sample. The reaction only takes place under specific conditions and with a specific form of the metal and therefore all the different forms of the metal present in the sample must be converted into the reactive form in order to ensure total metal concentration.

Colorimetric methods are fast and require little specialized apparatus therefore are usually the basis for field test measurements, emergency kits and for supporting the activities in the water treatment plants. These methods can be automated via an automatic analyser system working in a batch or in continuous flow mode such as flow injection analysis.

The applicability of colorimetric methods is limited by the chemical interferences occurring due to the presence of other reactive metal ions or substances, enhancing or suppressing the analyte concentration. It is necessary to perform matrix and interferences studies in the validation processes for each matrix that is intended to be analysed. High detection limits, single element analysis and low sample throughput are other disadvantages of this kind of methods (see Table 8.3).

Table 8.3 Colorimetric methods for iron and manganese analysis in water.

Organization	Method number	Name	Metal	Working range (in µg/l)
ISO	6332:1988	Water quality – Determination of iron – Spectrometric method using 1, 10-phenanthroline	Fe	10–5000
	6333:1986	Water quality – Determination of manganese – Formaldoxime spectrometric method	Mn	10–5000
SMEWW	3500 – Fe b	Iron – Phenanthroline method	Fe	5–400
	3500 – Mn b	Manganese – Persulfate method	Mn	5–1500
ASTM	D 1068d	Test method D – Photometric bathophenanthroline	Fe	40–1000

8.3 ELECTROCHEMICAL METHODS

Electrochemical based methods are based in the measurement of the potential and/or current in an electrochemical cell containing the analyte.

There are three basic types of those methods:

- Coulometry – based on the relationship of equivalence between total electric charge and chemical reaction.
- Potentiometry – based on the relationship between the concentrations of ions located in a solution and the electromotive force of an electrochemical cell in which this ion is one of the components. The potential of a solution between two electrodes is measured and is related to the concentration of the analyte. These methods usually use electrodes made selectively sensitive to the ion of interest.
- Voltammetry is based on the study of current vs. potential relationships. It uses a cell in which one electrode acts as a polarizable inert electrode that means changing its potential according to a special time program, in a certain potential range, and a non-polarizable reference electrode that means its potential remains constant in the course of the voltage change in the measuring cell. In the course of the potential change, components of the solution can be oxidized or reduced, producing a current that flows through the cell. This current is related with the concentration of the analyte. Polarography and stripping methods are included in this category.

Among the advantages of electrochemical methods are easy to use and non-expensive equipment. In voltammetry the detection limit for iron is 10^{-9} Mol/l. In some cases it is possible to perform speciation analysis, since this kind of method responds to the oxidation states of metals.

One of the most important disadvantages of these methods is the dependence of the matrix of the sample. Precise pH control is very important in order to get reliable results. Organic colour and presence of organic compounds in the water can cause interferences and in some cases, complex sample preparation may be required.

8.4 ATOMIC ABSORPTION SPECTROSCOPY

Atomic absorption spectroscopy uses an atomic absorption which occurs when a ground state atom absorbs energy in the form of light of a specific wavelength and is elevated to an excited state. The amount of light absorbed is related to the number of atoms of the selected element present in the light path. The relationship between the amount of light absorbed and the concentration of the selected element present in known standards can be used to determine unknown sample concentrations by measuring the amount of light absorbed. The instrumental equipment includes:

- Source of light, usually a hollow cathode lamp or electrode-less discharge lamp.
- Sample introduction device.
- Atom source (system to produce atoms in the elemental state).
- Monochromator to isolate specific wavelength.
- Detector to measure the amount of light accurately.
- Data processing system.

The main differences between particular atomic absorption spectroscopy methods are related to the sample introduction system and the atom source.

There are two types of atomic absorption spectroscopy suitable for determination of iron and manganese in water:

(1) Flame Atomic Absorption Spectroscopy (F AAS).
(2) Graphite Furnace Atomic Absorption Spectroscopy (GF AAS).

In F AAS the source of energy which produces free atoms is heat. Most commonly it is an air/acetylene or nitrous-oxide/acetylene flame. The sample is introduced into the flame, as an aerosol produced in the nebulizer and spray chamber system. The burner head is aligned so that the light beam passes through the flame and light could be absorbed by the free atoms of the sample.

The system is relatively robust and complex matrices can be analysed by this method. Analysis is fast (10–15 s per sample per element). It is low in cost and the measurement can easily be automated.

The major limitation is the relatively low efficiency of the burner–nebulizer system which causes a degradation of the detection limits.

In Graphite Furnace Atomic Absorption Spectroscopy (GF AAS) the sample is introduced directly into a graphite tube in the form of a single drop, and then the tube is heated in a programmed series of steps in order to eliminate the solvent, remove the major sample matrix components and to atomize the metals. The tube is aligned with the light path to allow absorption of the selected wavelength.

The advantage of this method is that it requires only small quantities of sample. Usually only 10–20 μl is injected for single metal. Sensitivity and detection limits are significantly improved compared to F AAS (see Table 8.4).

Table 8.4 Detection limits (in μg/l) for iron and manganese in F AAS and GF AAS methods.

	F AAS	GF AAS
Iron	5	0.03–0.05
Manganese	1.5	0.035–0.05

One of the main disadvantages of the GF AAS is the relatively small analytical working range (2–3 orders of magnitude of the signal intensity). Another one is that the temperature program should be adjusted to each different sample matrix. Analyses are slower than in F AAS method and costs per analysis are higher.

8.5 INDUCTIVELY COUPLED PLASMA SPECTROMETRY

Optical Emission ICP-OES

ICP-OES, also called ICP-AES, is based upon the measurement of the light emitted by the elements in a sample that was introduced in the plasma as an aerosol produced in a nebulizer/spray chamber system. The high temperature of the plasma supplies energy to vaporize solvent, eliminates sample matrix components and elevates atoms to their excited states. The intensity of the emitted light is related to the number of atoms in the plasma and to the concentration of the analyte in the sample. Each metal has a characteristic spectrum of emission lines that can be used for analytical purposes.

There are two ways of viewing the light emitted from an ICP:

- Radial configuration – the light across the plasma is viewed radially, resulting in the highest upper linear ranges and minimizing spectral interferences.
- Axial configuration – the light path is maximized and is on axis with the plasma. Detection limits can be enhanced by as much as a factor of 10 when compared with radial configuration but spectral interferences are much more difficult to eliminate.

There are instruments available on the market, capable of working with both configurations if required.

Mass Spectrometry ICP-MS

In ICP-MS, the argon plasma generates singly charged ions from the elemental species within a sample. The ions are directed into a mass spectrometer and separated according to their mass/charge ratio then directed to a detector that determines the number of ions present. Three different types of mass analysers have been used with ICP-MS which are:

- Quadrupole.
- Magnetic sector.
- Time-of-flight analysers.

The quadrupole mass spectrometer is the most commonly used because of its easy-to-use, robustness and working speed.

ICP-MS combines the multi-element capabilities of ICP techniques with detection extremely low detection limits (see Table 8.5) and analytical working range of nine orders of magnitude. ICP-MS instruments require computers and sophisticated software to control the mass spectrometer and to perform calculations on the data collected.

Table 8.5 Detection limits (in μg/l) for iron and manganese achieved by ICP-OES and ICP-MS.

	ICP-OES	ICP-MS
Iron	0.5–2	0.001–0.002
Manganese	0.4–1	0.0001–0.0005

Due to the working principle of ICP-MS several spectral interferences can be observed. Isobaric interferences are caused by isotopes of different metals that have the same mass/charge ratio. Polyatomic interferences are caused by ionic recombination which occurs in the interface region and produces charged molecules that could have the same mass/charge ratio of the metal of interest. Table 8.6 presents the most common interferences involving iron and manganese.

Table 8.6 Common interferences with iron and manganese in ICP-MS.

Polyatomic species	Analyte
$^{40}Ar^{16}O$	^{56}Fe
$^{40}Ar^{16}O^{1}H$	^{57}Fe
$^{38}Ar^{16}O^{1}H$	^{55}Mn

Some of these interferences can be corrected using analysis of different isotopes or using reaction/collision cell technologies. These technologies allow most of interferences to be removed before they reach the mass spectrometer by using controlled gas-phase reaction chemistry or kinetic energy inside an enclosed cell. In modern ICP-MS instruments, many correction equations are built into the instrument software to facilitate automatic correction of certain isobaric or polyatomic interferences.

8.6 SPECIATION ANALYSIS

Both iron and manganese normally exist in water as reduced Fe(II) and Mn(II) or oxidised, Fe(III) and Mn(IV) ions. Metal-organic complexes can also be present. Detailed knowledge about the form in which iron and manganese is present in raw water is essential for determining the most suitable treatment method.

Quite often, the speciation of iron and manganese is determined by physical separation using filtration through various pore size filters (0.1–0.45 micrometres) followed by ultrafiltration. The general assumption is that reduced iron and manganese are soluble and oxidized forms are not. Therefore, determinations on a filtered sample reflect concentrations of reduced species, and determinations on an unfiltered sample yield a measurement of oxidized (or reduced and oxidized) species, depending on analytical methods. While the use of filtration to differentiate oxidation states of iron and manganese is easily conducted in field studies, speciation of iron by filtration may not always be applicable.

Computer modeling to determine Fe and Mn speciation based on solubility relationships is possible, but may not bear resemblance to reality. Modeling is most useful when applied to speciation of metals under equilibrium conditions. Additional on-site measurements such as pH, redox potential and temperature, are required in order to obtain reliable results.

Iron

Traditional tests for distinguishing Fe(II) from Fe(III) include spectrochemical methods which generate a coloured complex with Fe(II). Typical complexing agents used include phenanthroline, bathophenanthroline, and ferrozine. Iron in water samples treated with one of these agents forms a red-orange (or purple) complex. This highly coloured species absorbs visible light, and thus its concentration can be determined by spectrophotometric analysis. Content of Fe(III) is then determined by difference between Fe(II) total and Fe which is determined (using different method) separately. Some of these methods are quite sensitive and allow determining iron speciation at the level of micrograms per litre. However, the tests are usually applied to the milligrams per litre range due to the problems with the stability of Fe(II) in the presence of atmospheric oxygen and light. As a result, preconcentration and separation are usually required before actual analysis. It has been also demonstrated that Fe(III) in solution can react with ferrozine, thereby interfering with the assay of Fe(II) (Gibbs 1976; Kieber *et al.* 2001).

Recently, speciation of Fe(II) and Fe(III) has been studied with different techniques such as ICP-AES and polarography, cathodic stripping voltammetry, capillary electrophoresis, flow injection chemiluminescence, flame AAS, flow injection analysis and automated analysis.

The most popular method is high performance liquid chromatography combined with inductively coupled plasma mass spectrometry method (HPLC/ICP-MS).

Manganese

Very few analytical techniques can measure the concentration of one particular species of manganese in solution, thus analytical techniques for chemical speciation of Mn are limited.

Anodic stripping voltammetry can be used to determine the complexation capacity of natural waters for manganese (Kalavska 1991; Florence 1982). This method distinguishes between "labile" and "bound" metal species. Following a sample titration, the bound metal is determined as the difference between total and labile metal and the strength of the bound metal-complexes can be determined mathematically. Then a chemical equilibrium/speciation computer program can be used to more specifically determine the Mn chemical speciation. The analytical technique itself, however, is not species-specific, and is subject to interference by organic compounds present in natural waters. Additionally it is time consuming (usually only one sample can be titrated per day).

Potentiometric stripping analysis and differential pulse polarography have been applied to determine Mn in natural waters, but these methods suffer from interferences and are limited in their ability to speciate Mn (Chiswell & Mokhtar 1986).

Ion exchange chelating columns have also been used to separate Mn(II)(aq) from colloidal Mn (Chiswell & Mokhtar 1986; Kalavska 1991).

An on-line sequential injection analysis system with spectrophotometric detection using 4-(2-pyridylazo) resorcinol as reagent was proposed for the speciation of Mn(II) and Mn(VII) by van Staden and co-workers. The system achieved detection limit of 0.005 mg/l for total Mn and 0.008 mg/l for Mn(II) with the sampling frequency of 30 determinations per hour (van Staden *et al.* 2003).

Physical separation techniques are used to separate "soluble" from "insoluble" forms of Mn. Some laboratories use a method based on the size of manganese-containing organic or colloidal molecules. Water samples are processed through a progression of membrane filters having different pore sizes down to a 500 molecular weight cut-off. The membrane filters are mounted into centrifuge tubes and placed into a high speed centrifuge. The collected filtrates from each different size of filter are then subsequently analysed by ICP-MS.

Fractional procedure described by Carlson *et al.* (1997) uses two filtration steps. The water sample is initially filtered through a 0.22 μm filter to remove particulate metal species. Then the filtrate is passed through an ultrafilter (usually 30,000-Dalton molecular weight cutoff) to separate the colloidal and dissolved species. The initial filtration separates out the particulate Mn and leaves a fraction that is called "dissolved". This fraction actually consists of both colloidal and "truly dissolved" Mn. After ultrafiltration "truly dissolved" Mn remains in the filtrate and it is assumed to be Mn(II).

KEY POINTS:

(1) Selecting the most appropriate method is the key for achieving accurate and reliable results.
(2) For the determination of iron and manganese in water various types of methods can be used: colorimetric methods, electrochemistry methods, atomic absorption spectroscopy, inductively coupled plasma.
(3) Each analytical technique has its own advantages and limitations.
(4) Legal requirements may sometimes determine the choice of laboratory method, but in most cases only the performance characteristics are specified.
(5) Detailed knowledge about the form in which iron and manganese is present in water is essential for determining the most suitable treatment method.
(6) Computer modeling to determine Fe and Mn speciation based on solubility relationships is possible, but may not bear resemblance to reality. Modeling presents reliable results when applied to speciation of metals under equilibrium conditions.

Chapter 9

Investigational methods

9.1 DIRECT CORROSION TESTING

In all corrosion testing procedures, the one which most nearly duplicates the conditions in distribution network, will provide the most satisfactory and reliable results. In numerous cases corrosion test results obtained for specific set of conditions are extrapolated to completely different situations. However, results published without data on time of exposure, stress, stray currents, composition of the metal or alloy, the quality of the water environment (pH, temperature, dissolved oxygen, mineral analysis), concentration of inhibitor, rate of flow, and a host of other variables, are unreliable and may lead to drawing completely wrong conclusions.

It is necessary to conduct laboratory studies under controlled conditions to evaluate the relative effects of primary variables as a basis for extrapolation for specific applications. Simulating the normal (real life) corrosion rate of metals is very important. By design, metals in normal conditions are exposed for many years. The time element in laboratory studies is much shorter. It is well known that at first exposure, metals in laboratory experiments corrode at a very high rate. After the first exposure, corrosion rate decreases to a constant or equilibrium rate which may take 50 to more than 100 days. This equilibrium rate is the one closest to normal exposure.

Coupons and pipe rig systems are good tools to compare different corrosion control techniques prior to initiation of system-wide corrosion control programs. They provide a viable means to simulate distribution systems without affecting the integrity of the full scale system. Even with a prolonged conditioning period for the materials in the water of interest, coupons used in the field or laboratory and pipe rig systems cannot give an exact assessment of the corrosion of larger distribution systems since too many factors influence water chemical composition at the consumer's tap.

The corrosion rate (CR) can be calculated using formula:

$$\text{Corrosion Rate} = \frac{\text{Weight loss} \times \text{K}}{\text{Metal Density} \times \text{Exposed Area} \times \text{Exposure Time}}$$

where:

K is a density factor.
K = 1 for carbonate steel.

Corrosion rates can be expressed in various units. The most commonly used are millimetres per year (mm/y) and mils (0.00 1inch) penetration per year (mpy). Unit conversion tables can be found in many publications (e.g. Shreir *et al.* 1994; Baboian 2002, 2005). Corrosion rates calculators are also easily available via the internet (e.g. http://corrosionsource.com/ or http://metals.about.com/od/Corrosion/a/Corrosion-Rate-Calculator.htm).

Corrosion scale and metal surface evaluation

Direct examination of corrosion products and corroded pipes inner surfaces is one of the most popular and important methods of corrosion investigation methods. Understanding of corrosion processes often requires detailed knowledge of the morphology, composition and structure of corroded surface and near surface regions of a pipe as well as corrosion deposits and bulk water. Due to complexity of chemical and biochemical processes that occur in distribution system a combination of various techniques is required to gain sufficient information.

The first step in corrosion scale examination is macroscopic observation of scale on the inside of the pipe. If a pipe is free of scales, an examination of the pipe wall can yield information about the type and extent of corrosion and corrosion-product formation (such as tubercles).

Examination under a microscope can provide more information, such as local corrosion too small to be seen by the unaided eye. Photographs of specimens should be taken for comparison with future visual examinations. High magnification photographs should be taken, if possible.

When more detailed information about corrosion products structure and chemical composition are needed, additional techniques can be used such as: atomic force microscopy, scanning electron microscopy, energy-dispersive spectroscopy, time-of-flight secondary ion mass spectrometry, X-ray powder diffractometry, grazing incident diffractometry, transmission electron microscopy, selected area electron diffraction, X-ray photoelectron spectroscopy, Fourier transform infrared, Mossbauer spectroscopy, Raman spectroscopy. Usefulness of various techniques that can be used in investigation of iron pipe corrosion products is discussed in numerous publications (e.g. Borch *et al.* 2008).

Microbially influenced corrosion is a very important problem in many distribution systems. Iron bacteria are well associated with this type of corrosion. When the equilibrium of regrowth processes is disturbed or hindered, bacteria may be released into the bulk water causing additional problems (LeChevallier *et al.* 1987).

Investigations of biofilms bacteria present in loose deposits and bulk water are conducted using standard microbiological techniques (Charaklis 1990; Appenzeller *et al.* 2001).

Pipe rig systems

Pipe rig system can be designed to capture several of the effects of various water quality parameters (such as pH, DO, alkalinity etc.) that influence the corrosion of iron. In addition to measuring mass loss rate per unit of surface area it can be used to determine the corrosion rate. Furthermore, pipe rig systems can simulate a distribution system and/or plumbing system and allow for the measurement of certain constituents, depending on which corrosion control strategy is used.

The selection of the most appropriate materials for the conditions under study is critical to achieve the most reliable approximation. These systems, which can be made from new materials or sections of existing pipes, are conditioned to allow for the development of corrosion scales or passivating films that influence both the corrosion rate of the underlying metal and the metal release. When using harvested pipe segments, it is important not to disturb existing pipe scales and to run the loop at baseline conditions for sufficient time. The conditioning period must allow for the development of corrosion scales and allow the pipe scale to stabilize, which may vary from weeks to years, depending on the type of metal examined (AWWARF 1996). Owing to this variability, 6 months is recommended as the minimum study duration (Eisnor & Gagnon 2003).

As with coupon testing, there is currently no single standard for the use of pipe rig systems in the evaluation of corrosion in drinking water distribution systems. Several authors have presented various systems used in corrosion studies (e.g. Prakash *et al.* 1988; AWWARF 1990, 1996; Holden *et al.* 1995; Smith *et al.* 1999; Sarin 2002; Imran *et al.* 2005a; Lytle *et al.* 2005; Tang *et al.* 2006; Husband *et al.* 2008). Figure 9.1 presents generalised schematics of pipe rig system that can be used for corrosion testing and examination of an influence of different chemicals additions (e.g. inhibitors) on corrosion rates.

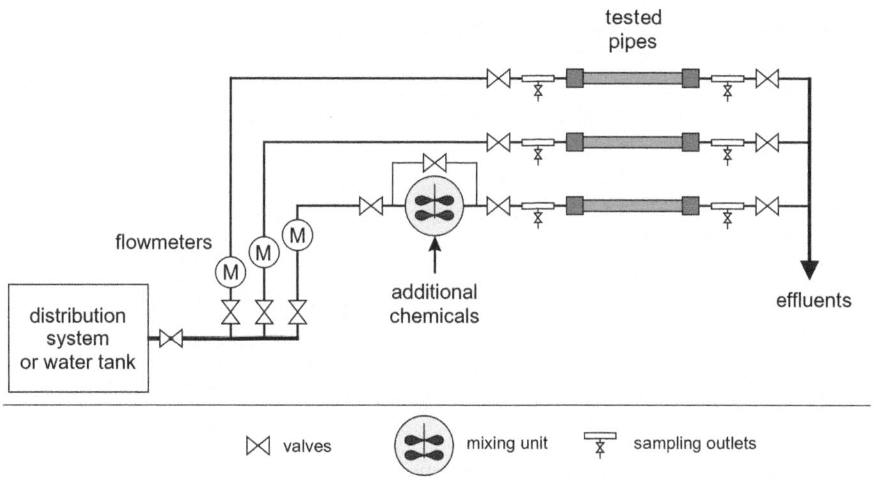

Figure 9.1 Generalised schematics of pipe testing setup.

Eisnor and Gagnon (2003) proposed a framework for the implementation and design of pilot-scale pipe testing setups. This framework identified eight important factors that should be taken into consideration when designing pipe rig systems:

(1) Test section style (permanent or inserts).
(2) Test section materials.
(3) Test section diameter.
(4) Test section length.
(5) Flow configuration.
(6) Retention time.
(7) Velocity.
(8) Stagnation time.

Metal coupons

Coupons inserted in the distribution system are typically used to determine the corrosion rate associated with a specific metal. They provide a good estimate of the corrosion rate and allow for visual evidence of the scale morphology. Weight loss measurement is still the most widely used means of determining corrosion rate. It is simple and applicable to all corrosive environments, and gives information on effects of all forms of corrosion (Williams *et al.* 1984).

The coupon metal used must be representative of the piping material under investigation. The coupons are typically inserted in the distribution system for a fixed period of time, and the corrosion rate is determined by measuring the mass loss rate per unit of surface area. The duration of the test must allow for the development of corrosion scales, which may vary from single weeks to 24 months, depending on the type of material examined.

There is currently no single standard regarding coupon geometry, materials or exposure protocols in drinking water systems.

Commercially available corrosion coupon racks are usually made from plastic materials such as PVC and allow simultaneous testing of 2–8 coupons of rectangular shape (Figure 9.2). Typical coupon dimensions are: length 75–80 mm, width about 13 mm and thickness 1–2 mm.

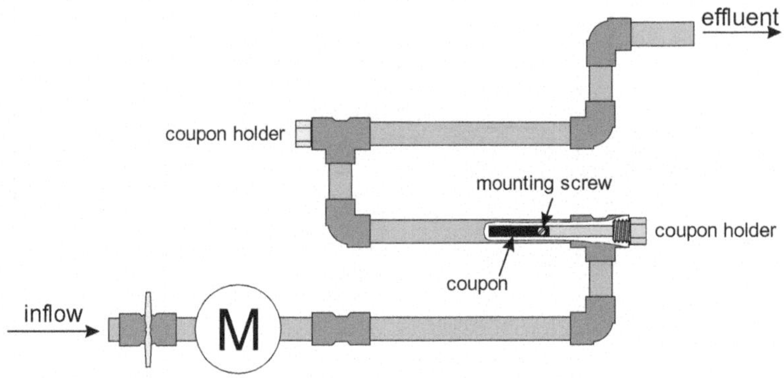

Figure 9.2 Schematic diagram of popular corrosion coupon rack.

Another method of coupon testing can be used in large diameter pipes using fixed or retractable rod-shaped coupon holders. The retractable coupon holder is contained in the retraction chamber which is bolted to the gate valve thus allows the coupons to be slid into the pipe. The retraction chamber contains a packing gland through which the coupon holder can pass. Coupons mounted on the rod in the extended position are drawn into the retraction chamber.

The major drawback of coupons is their poor reproducibility performance due to high degree of variation between individual coupon measurements. This lack of precision is due both to the complex sequence of handling, preparation and surface restoration procedures, which provides opportunity for analysis-induced errors, and to the high degree of variability that exists in metallurgical properties or chemical conditions on the coupon surface during exposure (Reiber *et al.* 1988, 1996). Procedures for mass-loss coupon analysis can be quite labour intensive. Achieved results may be misleading in situations where the corrosion rate varies significantly over time because of unrealized processes and factors (Roberge 2007).

Immersion tests

There are many variations in a laboratory immersion test (Guthrie *et al.* 2002). Historically, immersion tests have been extensively used to obtain uniform corrosion data. General guides on how these tests may be performed are described in standards issued by American Society for Testing and Materials (ASTM G31-72 2004) and the National Association of Corrosion Engineers, NACE TMO 169 (Baboian 2005).

Basically, small pieces (e.g. 4-cm by 7-cm by 0.3-cm) of the tested material are exposed to the water (natural or with added constituents) and the loss of weight of the material is measured for a given period of time. In most simplified version the metal sample is placed in a jar containing water or test solution, thus the term "jar tests".

According to ASTM G31 an apparatus should be used, consisting of a kettle or flask of suitable size (usually 500 to 5000 ml), a reflux condenser with atmospheric seal, a sparger for controlling aeration a and temperature regulating device, a heating device, and a coupon support system. If agitation is required, the apparatus can be modified to accept a suitable stirring mechanism. One of the critical parts of corrosion testing is the preparation of the specimen to be tested. It is desirable to use the same surface conditions for the tests as those found in service. This is not always possible because of the variability of the material's surface in service and their age.

Immersion testing is a good method for screening studies. This technique is often the quickest and most economical means for providing a preliminary estimation of potential corrosion problems (Letterman 1999). Besides the full immersion test, there are several variations that could be used to accentuate certain types of exposure conditions.

The most serious disadvantage of this method is the assumed average-time weight loss (Reiber *et al.* 1988). The corrosion rate could be high initially and then decrease with time. In other cases, the rate of corrosion might increase very gradually with time, it could cycle, or it could be some combination of these things. A survey performed by Volk and co-workers (Volk *et al.* 2000) on cast iron, showed significant changes in corrosion ratios over time. During 3 months experiment corrosion rates varied from below 3 mpy up to 7 mpy, even when a corrosion inhibitors (at constant doses of 0.86 mg PO_4/l), were used. It was also presented that corrosion rates were strongly related to water temperature.

Electrochemical measurements

It was observed that during polarization tests at low applications of current to the specimens, the change in the potential of the specimens was high for high corrosion rates, and low for low corrosion rates. T. E. Larson and R. V. Skold have developed an empirical relationship between the instantaneous corrosion rate and the change in potential per unit of applied current density (Larson & King 1954; Larson & Skold 1957, 1958). Within a few years after the publication of this discovery the theory and improved application of this technique were developed by many investigators. More detailed descriptions on using these methods can be found in several publications concerning corrosion (e.g. Mansfeld 1976; US EPA 1982; Letterman 1999; Scully 2000; Frankel 2008; Revie & Uhlig 2008; Cottis *et al.* 2010).

Corrosion testing with polarization methods is based on forcing potential or current changes on a sample under study while monitoring the resulting response in current or potential. Polarization resistance technique is very well established for routine use both for corrosion prediction and corrosion monitoring (Mansfeld 1976; Scully 2000). During routine use, the realistic limit for corrosion rate estimation is about 0.001 mm/year (Revie 2000). Lee and co-workers used polarization method during a study on effect of nitrite on the corrosion behaviour of carbon steel pipelines. They measured corrosion rates ranging from 0.00171 to 0.171 mm/year (Lee *et al.* 2012).

9.2 CORROSIVE PROPERTIES OF WATER – STABILITY INDICES

The most important water quality parameters affecting corrosivity are believed to be: alkalinity, pH and calcium content. In drinking water supply, chlorides and sulphates also play an important role.

Although a number of indices have been developed, none has demonstrated the ability to accurately quantify and predict the corrosivity or aggressiveness of water. The indices can only give a probable indication of the potential corrosivity of water. Experience has shown that if conditions encourage the formation of a protective calcium carbonate film, then corrosion will generally be minimized. Several models and indices are available that use calcium carbonate chemistry to evaluate water stability but are not directly applicable when assessing corrosivity of water towards metals such as cast iron or steel.

The nature of the material is important since biological growths in the distribution system can also cause corrosion by providing an environment in which physical and chemical interactions can occur.

Organic matter found in water can significantly influence corrosive properties of water. Humic and fulvic substances are major components of the natural organic matter. Several authors have suggested decreases iron corrosion rate of galvanized steel and cast iron (Larson, 1966; Sontheimer *et al.* 1981; Sander *et al.* 1997; Broo *et al.* 1999). However, other experiments produced different results: presence of natural organic matter increased the corrosion rate at low pH values, but decreased it at high pH values due to different surface complexation (Broo *et al.* 2001; Pullin & Cabaniss 2003). Therefore water stability and corrosion indices will not provide reliable results when dealing with water containing fulvic and humic substances.

Langelier saturation index (LSI)

Wilhelm Langelier (1936) developed a method for predicting the pH at which water is saturated in calcium carbonate (called pH_s). This index provides a measure of the stability of water with respect to its degree of $CaCO_3$ saturation. The pH at which water is saturated with $CaCO_3$ is known as the pH of saturation or pH_s.

$$LSI = pH_{measured} - pH_s$$

Saturated water has a LSI of zero. If water has a negative LSI value, it is under-saturated with respect to calcium carbonate which means that the water is potentially corrosive. For waters with a positive LSI, a protective layer of calcium carbonate can form as the water is oversaturated with $CaCO_3$ and the water is scaling. In practice, water is considered to be potentially aggressive if it has a LSI of less than –1.5 (Gebbie 2000).

The LSI describes only the corrosivity of water towards an existing calcium carbonate scale, or other calcium carbonate bearing structure. The interpretation of corrosivity towards metals is not explicit in the LSI. It was described in numerous studies that waters supersaturated with calcium carbonate often develop tubercular deposit which do not inhibit corrosion on mild steel (Stumm 1986; Loewenthal & Marais 1982).

It should be mentioned that LSI is temperature sensitive. The LSI becomes more positive as the water temperature increases.

There are many LSI calculators available via the internet that allow calculating LSI index by simply inputting basic water characteristics such as: pH, temperature, total dissolved solids, Ca and alkalinity (e.g. http://www.lenntech.com/calculators/langelier/index/langelier.htm; http://www.advanced-water-systems.com/technical/calculations_and_tables/langlier.html).

Ryznar index (RSI)

This saturation index was developed in 1944 by John W. Ryznar (Ryznar 1944) from empirical observations of corrosion rates and film formation in steel mains. RSI takes the form:

$$RSI = 2(pH_s) - pH$$

Unlike in case of LSI, there is no theoretical basis for RSI. It was based solely upon scaling performance observed in a large number of water samples. For those waters, it was seen that:

- $RSI < 5.5$ – heavy scale will form.
- $5.5 < RSI < 6.2$ – scale will form.
- $6.2 < RSI < 6.8$ – no difficulties. Scaling is unlikely to occur.
- $6.8 < RSI < 8.5$ – water is aggressive.
- $RSI > 8.5$ – water is very aggressive.

It should be noted that due to the empirical nature of this relationship, when it applies to a large number of water samples, there may also be many water samples for which it does not apply.

Larson-Skold index (LR)

T. E. Larson and R. V. Skold developed a simple index in a study to quantify the aggressiveness of water toward mild steel and cast iron (Larson & Skold 1957, 1958). It was based upon chloride and sulphate being the major contributors to the aggressiveness toward corrosion and alkalinity as working to minimize their aggressiveness. The study was conducted on Great Lakes waters and tends to have some ability to predict the likelihood of pitting on waters of similar composition.

The index is the ratio of equivalents per million (epm) of sulphate (SO_4) and chloride (Cl) to the equivalents per million of alkalinity in the form bicarbonate plus carbonate ($HCO_3 + CO_3$).

$$LR = (\text{epm Cl} + \text{epm } SO_4)/(\text{epm } HCO_3 + \text{epm } CO_3)$$

The interpretation of Larson-Skold index might be made using following guidelines:

- $LR \ll 0.2$ – chlorides and sulphate probably will not interfere with natural film formation.
- $LR \gg 0.2$ but $\ll 0.6$ – chlorides and sulphates may interfere with natural film formation; higher corrosion rates might be anticipated.
- $LR \gg 0.6$ – the tendency towards high corrosion rates of a local type should be expected as the index increases.

Since the Larson-Skold relationship is based upon Great Lakes waters, extrapolation to other waters, such as those of low alkalinity or very high alkalinity, goes beyond the range of the original data. The extrapolations should be closely scrutinized or weighted lightly. It should be mentioned that the waters studied by Larson and Skold were capable of forming an inhibitory calcium carbonate film, if no interferences were present. Chlorides and sulphates increase the aggressiveness of water with adequate buffering capacity and alkalinity to otherwise not be overly aggressive.

Imran and co-workers (Imran *et al.* 2005) evaluated many corrosion indices with data collected from a pilot distribution system. They found that the Larson Ratio had a better correlation to iron and colour release than most of the calcium carbonate based indices. They suggested modification to the Larson-Skold index Ratio (MLR) to account for the effects of sodium, temperature and hydraulic retention time (HRT) on iron release and to achieve a better correlation to the iron release data:

$$MLR = \frac{(Cl^- + SO_4{}^{2-} + NA^+)^{1/2}}{ALK} \frac{T}{25} HRT$$

where:

Cl^- is the concentration of chlorides in mg/l
$SO_4{}^{2-}$ is the concentration of sulfate in mg/l

Na^+ is the concentration of sodium in mg/l
T is the temperature in °C
HRT is the hydraulic retention time in days
ALK is the alkalinity of the sample in mg/l $CaCO_3$

Waters with a MLR value below 0.5 are considered noncorrosive and those with a value above are corrosive. The term "corrosive" in this context refers to the tendency of the source waters to cause iron based discolouration problems.

9.3 CONSUMER COMPLAINT LOGS

Complaints by customers are typically the first indication of an iron or/and manganese problem. However, customer complaints can arise for a number of reasons; aesthetic issues are usually the most common. If good records of complaints are kept, trends can be seen that may indicate a corrosion problem.

A complaint map should show the water distribution system with codes to indicate:

- The location.
- Frequency.
- The type of complaints received.

Using such a map it should be easy to determinate if a problem exists and where it may be located. As an alternative, customer surveys may also be used to get the same information more quickly than compiling records of customer complaints. It is very important to remember that sediment causing discolouration may have originated upstream rather than at the location of the complaints.

These methods are relatively inexpensive and, depending on how the data is used, can be very effective as a first step in a corrosion or metal control program.

9.4 OTHER TECHNIQUES

Aesthetic quality of water, especially colour, strongly affects consumer perception of water safety (see Chapter 6). Discolouration is caused by the presence of dissolved colloidal or suspended substances. The particles causing discolouration are predominately less than 10 micrometres with a specific gravity in the range 1 and 1.3 typically containing iron and manganese (Boxall *et al.* 2001; Gauthier *et al.* 2001; Sarin 2002; Sarin *et al.* 2004a; Seth *et al.* 2003; Vreeburg 2007).

Several researchers suggest that because of particulate nature and consistent fine size of the material that causes discolouration, water turbidity can be used as a measure of discolouration (e.g. Twort *et al.* 2000; Sarin 2002; Vreeburg & Boxall 2007; Husband *et al.* 2008; Husband & Boxall 2011). Turbidity is easily measurable optically by detecting light scattered on fine particles in the sample. Wide range of commercially available turbidimeters can be used for this purpose.

Another method for quantifying iron discolouration was developed in the UK by Water Research Center. The technique is based upon an automated filtration samplers deployed at strategic locations to investigate sections of water mains. The colour of filter pads used to filter set volumes of water can indicate the internal condition of pipework (Evins *et al.* 1990).

Controlled flushing via hydrants and washouts can also be used to estimate the quantities and investigate the composition of loose deposits (Evins *et al.* 1990; Friedman *et al.* 2002; Barbeau *et al.* 2005; Carriere *et al.* 2005).

KEY POINTS:

(1) The simulation of corrosion processes in pipe rigs or with coupon testing gives a direct measure of corrosion rate but due to the complicated nature of corrosion it is expensive and time consuming.
(2) Immersion testing is a good method for screening studies. This technique is often the quickest and most economical means for providing a preliminary estimation of potential corrosion problems.
(3) When using stability indices for the assessment of potential iron and manganese problems caused by corrosion one should remember that due to the empirical nature of those relationships when they apply to a large number of water supply systems, there may also be many systems for which they do not apply, especially when dealing with water containing fulvic and humic substances.
(4) The indices can only give a probable indication of the potential corrosivity of water.
(5) Consumer complains logs and consumer surveys can be very effective as a first step in a corrosion or metal control program.

Chapter 10

Corrective water treatment and other engineering methods

10.1 REASONS FOR CORRECTIVE TREATMENT

Iron and manganese, even at concentrations below WHO guide values may be a cause of problems for operators of water supply systems.

The most common problems are:

- Precipitation of sediments in pipelines.
- Formation of stains on wash-basins, bathtubs and washed clothes.
- Development of ferric and manganese bacteria in a distribution network which leads to pipeline corrosion and further microbial growth.
- Metallic taste of water.
- Deterioration of water odour.
- Increase in water colour.

Sediments and biofilm decrease the pipe diameter leading to an increase of hydraulic losses and energy costs in water distribution (Larson 1960). Bacterial and chemical corrosion significantly reduces the degree of reliability of a water distribution network causing an increase in the cost of repairs. This corrosion also results in deterioration of the quality of water delivered to consumers. These phenomena commonly occur at iron and manganese concentrations exceeding 0.3 mg Fe/l and 0.1 mg Mn/l, respectively. However, these concentrations do not protect water distribution network from accumulative sediments. Anglian Water in the UK set action limits for their potable water treatment works outputs of Iron 0.05 mg/l and Mn 0.02 mg/l. These figures represented the performance achievable by a well-designed and operated iron and manganese removal plant and served to minimise dirty water problems.

10.2 MECHANISM OF IRON AND MANGANESE REMOVAL FROM WATER

Iron and manganese are relatively soluble in a reducing environment such as low-oxygen-containing groundwater and low-oxygen surface water such as the hypo-limnetic waters of eutrophic lakes (MWH 2005).

Iron and manganese removal from groundwater is based on precipitation of hydrated iron oxides $Fe_2O_3 \cdot nH_2O$ and manganese oxides $MnO_2 \cdot mH_2O$; being practically insoluble these can be removed from water by filtration. Therefore the classical technology of groundwater treatment is based on the transformation of Fe(II) and Mn(II) compounds to the form of insoluble oxides of Fe(III) and Mn(IV).

Two principal methods of this transformation are (Figure 10.1):

- An increase of the oxidation potential of water solution by the use of oxidants
- An increase of pH of water solution by the dosing of bases, for example, calcium hydroxide.

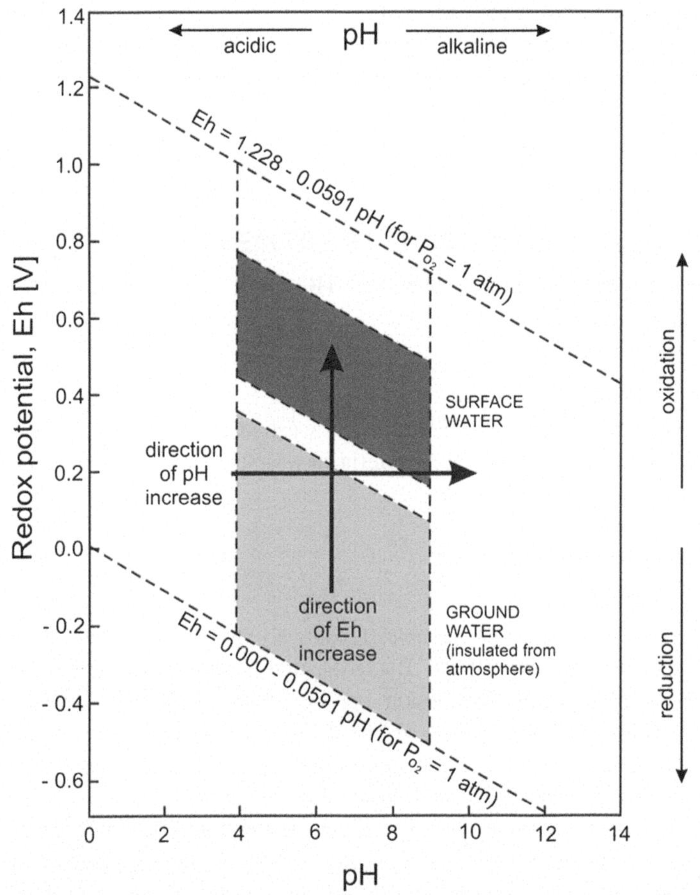

Figure 10.1 Principles of iron and manganese removal from treated water.

The oxidation reactions are very important for iron and manganese removal from water. Removal of manganese requires oxidation of Mn (II) to Mn(IV) (oxidation by two degrees of valence) and a higher redox potential. Thus manganese removal process is technologically more difficult than iron removal which require oxidation by one degree of valence: Fe(II) to Fe(III).

The basic oxidant used in iron and manganese removal process is oxygen from the air. Iron oxidation by oxygen consists of several steps defined by the following reactions (Faust & Aly 1997):

$$Fe^{+2} + O_{2(g)} \Longleftrightarrow Fe^{+3} + HO_2^{\bullet}$$

$$Fe^{+2} + HO_2^{\bullet} \Longleftrightarrow Fe^{+3} + H_2O_2$$

$$Fe^{+2} + H_2O_2 \Longleftrightarrow Fe^{+3} + HO^{\bullet} + H_2O$$

$$Fe^{+} + HO^{\bullet} \Longleftrightarrow Fe^{+3} + H_2O$$

Oxidation of manganese with oxygen (Morgan 1967) is a slow process in water of pH lower than 9.5. Oxidation of Mn(II) with oxygen does not proceed according to the law of first grade reaction, as it is observed in case of Fe(II) oxidation with oxygen. Mn(II) oxidation with oxygen is an autocatalytic process, with probable course defined by the equations:

$$Mn^{+2} \Longleftrightarrow + O_{2(g)}$$

$$Mn^{+2} \Longleftrightarrow + MnO_{2(s)}$$

$$Mn^{+2} \cdot MnO_{2(s)} + O_2 \Longleftrightarrow 2MnO_{2(s)}$$

Reactions of oxidation with oxygen proceed slowly, whereas the reaction in which Mn(II) is absorbed by MnO_2 proceeds more rapidly. Then, the adsorbed ions are slowly oxidized with oxygen present in the water (Benefield *et al.* 1982; Morgan 1967a, 1967b). The actual amount of oxidants necessary for manganese oxidation is smaller than the amount calculated from stoichiometric equation. The process of manganese removal is accelerated in the presence of MnO_2 and may be finished without complete oxidation of Mn(II).

In order to intensify the process of Fe(II) and Mn(II) oxidation, oxidants alternative to oxygen are used, for example: chlorine, chlorine dioxide, potassium permanganate and ozone. The reactions of Fe(II) and Mn (II) oxidation with the above oxidants are shown in Table 10.1.

Table 10.1 Oxidation reactions of Fe(II) and Mn(II).

Oxidant	Reaction
Iron	
O_3	$2Fe^{2+} + O_{3(aq)} + 5H_2O \longrightarrow 2Fe(OH)_{3(s)} + O_{2(aq)} + 4H^+$
HOCl	$2Fe^{2+} + HOCl + 5H_2O \longrightarrow 2Fe(OH)_{3(s)} + Cl^- + 5H^+$
ClO_2	$Fe^{2+} + ClO_2 + 3H_2O \longrightarrow Fe(OH)_{3(s)} + ClO_2^- + 3H^+$
$KMnO_4$	$3Fe^{2+} + MnO_4^- + 2H_2O \longrightarrow 3Fe(OH)_{3(s)} + MnO_{2(s)} + 5H^+$
Manganese	
O_3	$Mn^{2+} + O_{3(aq)} + H_2O \longrightarrow MnO_{2(s)} + O_{2(aq)} + H^+$
HOCl	$Mn^{2+} + HOCl + H_2O \longrightarrow MnO_{2(s)} + Cl^- + 3H^+$
ClO_2	$Mn^{2+} + ClO_2 + 2H_2O \longrightarrow MnO_{2(s)} + 2ClO_2^- + 4H^+$
$KMnO_4$	$3Mn^{2+} + 2MnO_4^- + 2H_2O \longrightarrow 5MnO_{2(s)} + 4H^+$

Table 10.2 presents stoichiometric values concerning the required amount of oxidant in milligrams of oxidant per milligram of Fe(II) or Mn(II) to be oxidized.

Table 10.2 Stoichiometric oxidant demand for Fe(II) and Mn(II) oxidation.

Element	Oxidant	Stoichiometric oxidant demand [mg of oxidant/mg of Fe or Mn]
Iron	O_3	0.43
	Cl_2	0.64
	ClO_2	1.20
	$KMnO_4$	0.94
Manganese	O_3	0.88
	Cl_2	1.29
	ClO_2	2.45
	$KMnO_4$	1.92

10.3 TECHNOLOGY OF IRON AND MANGANESE REMOVAL FROM WATER

Methods for iron and manganese oxidation

Aeration and chemical oxidation

Raw water aeration allows:

- Water oxygenation, an increase of redox potential and a change of the environment from reductive to oxidative.
- Removal of aggressive carbon dioxide.
- Removal of other gases from water, for example hydrogen sulphide.

There are a number of devices for raw water aeration. The basic criteria for choosing the aeration device are the alkalinity value of the raw water and the degree of aeration required. Practically, criterion values of alkalinity and corresponding devices are defined as follows:

- Alkalinity above 250 mg $CaCO_3$/l – pressure aerators.
- Alkalinity in the range of 150–250 mg $CaCO_3$/l – open aeration, spraying with the use of aerating nozzles (Figure 10.2), or open aeration in cascade aerators (Figure 10.3). The nozzles are installed on horizontally located distributing pipelines, to which water is supplied under pressure and due to installed nozzles; it is sprayed into vertical streams. Water aerated in this way is delivered to a reaction chamber (retention) of volume ensuring retention time determined by the kinetics of Fe (II) oxidation and flocculation of Fe_3O_3 nH_2O.
- Alkalinity in the range of 150–250 mg $CaCO_3$/l open aeration: multiple tray aerators, forced-draft tower aerator, natural draft tower aerator, packed tower aerator. Under natural or forced-draft tower aerators the process of oxygen absorption and removal of CO_2 from water occurs. Filling of shelves and towers uses different kinds of materials (e.g. basaltic, granitic broken stone, coke, boulders) and ceramic and plastic profiles (Rashing rings, Berl saddles, etc.). The purpose of the filling is to increase the surface of separation of water-air phases. The air flows from the bottom to the top in a counter-current direction in relation to water. Air flow may be caused by water flow or ventilator action. The filling material becomes coated with time with ferric oxide which intensifies the oxidation of Fe(II). After aeration the water may be delivered to a reaction tank and further on to rapid filters or to sedimentation tanks and then filters.

Figure 10.2 Open aeration; aerating nozzles [photograph courtesy of Paweł Pruss].

Figure 10.3 Open aeration in cascade aerators [photograph. courtesy of Paweł Pruss].

When alkalinity of raw water is lower than 150 mg $CaCO_3/l$, the removal of carbon dioxide requires chemical binding for example, with lime.

For hard waters, where the subsequent filtration is by rapid gravity filtration (open to the atmosphere) rather than pressure filtration, it is common practice to restrict the degree of aeration such as to oxidise iron and possibly manganese but avoid removal of carbon dioxide. This restricts the rise in pH and consequent precipitation of calcium carbonate during and after treatment. A minimum dissolved oxygen level of 60% saturation is usually adequate. This degree of aeration can often be achieved by the use of a simple single stage cascade aerator.

After the aeration, water is directed to rapid pressure or open-gravity filters (Figure 10.4). The water is filtered through filtration material of an appropriate granulation (0.7–2.5 mm). Filtration materials often used for iron and manganese removal are: silica sand, manganese ores, anthracite, green sand or materials produced according to patented technologies (Sommerfield 1999). Filtration of groundwater proceeds usually with a filtration rate from 7 to 12 m/h.

Figure 10.4 Rapid filters [photograph courtesy of Paweł Pruss].

In small plants for groundwater treatment, usually pressure filtration is used, whereas the composition of the raw water (mainly alkalinity) influences the choice of aeration device. Pressure aeration is preferred, as it decreases the number of water pumping stages. In the pressure system, water before filtration is carefully degassed. In the beds of pressure filters, manganese and iron are removed as a result of catalytic oxidation.

A modification of these technologies is a system with the use of coagulation, in which, after aeration, coagulant is dosed and water is directed to a suspended solids sedimentation tank. This system is used in the treatment of water with high colour and oxygen consumption and of high iron content found in complexes with humic and fulvic acids. This system is recommended for high capacity water treatment plants due to the complex operation of the processes. Another recommended technology of water treatment involving stable complexes of iron and manganese with organic compounds (e.g. waters from Miocene sediments) is contact coagulation or filtration aided with coagulation (Perchuć 1996).

The alternative for coagulation may be a decarbonisation process used for water of high carbonate hardness, reduction of which is one of the purposes of water treatment. Before sludge blanked clarification alkalis are dosed. The effects of removal of iron and manganese in this system increases along with a rise in water pH. Suspended sludge in a clarifier composed of hydrogenated and catalytic ferric oxides, intensifies Fe(II) oxidation.

For the intensification of Fe(II) and Mn(II) oxidation, especially when iron and manganese are present in complexes with organic compounds, chemical oxidation is used (Cl_2, O_3, $KMnO_4$, …) after the aeration process. Where chlorine is used, care must be taken to avoid breaching trihalomethane (THM) upper limits in the treated water at the customer's tap.

Microbiological oxidation

Oxidation of iron and manganese may proceed by chemical or biochemical methods. The activity of iron and manganese-iron bacteria may intensify and accelerate the processes of oxidation of Fe(II) and Mn(II). These are chemolitotrophic bacteria, which use energy from the oxidation of iron and manganese for assimilation of CO_2 (Olańczuk-Neyman 1988). These processes are used for groundwater treatment.

In groundwater with a low level of nutrients, the most commonly found bacteria are of genus *Achromobacter, Flavobacterium, Micrococcus, Nocardia* and *Cytophaga* and iron (*Sphaerotilus, Leptothrix, Acidithiobacillus*), sulfuric (*Thiothrix, Beggiatoa*) and manganese (*Leptothrix, Metallogenium*) bacteria. All these microorganisms carry out also respiratory processes which are very important for the technology of water production.

Bacteria may play the most important role in groundwater treatment for iron and manganese and are also very important in distribution. Their basic characteristics are shown in Table 10.3.

Table 10.3 Aerobic chemolitotrophic bacteria related to iron and manganese metabolism (using inorganic compounds).

Inorganic donor of electrons	Oxidation product	Group of bacteria	Example of bacterial species
Fe(II)	Fe(III)	Iron bacteria	*Acidithiobacillus ferrooxidans*
Mn(II)	MnO_2	Manganese bacteria	*Leptothrix echinata*

Iron bacteria (ferric) are found in waters in which concentration of divalent iron is in the range of 0.15 to 8.5 mg/l. These bacteria oxidize Fe(II) to Fe(III) and overgrow underwater objects and may clog up water pipes and cause their corrosion. Filiform iron bacteria which form a characteristic biofilm filled with gelatinous substance are best known.

Iron bacteria are a quite diversified group of microorganisms, because apart from iron oxidation, they can oxidize manganese at the same time. Due to this, they are divided into various groups (Bobrowski 2002).

Chemolithotrophic oxidizing bacteria – binding iron, but not oxidizing manganese; they are able to oxidize Fe(II) in aerobic conditions, and they use the obtained energy in the process of CO_2 assimilation. To this group belong *Acidithiobacillus ferrooxidans, Leptospirillum ferrooxidans* and species of the genus *Gallionella*.

Acidithiobacillus ferrooxidans are bacteria which grow in an acidic environment at pH 2.0–2.5 and a temperature of about 30°C. They synthesize organic matter from CO_2 and require much energetic input. Besides iron, they also use sulphur as an electron donor. Oxidation of sulphide or iron ions is a process of low energetic efficiency; therefore these bacteria must oxidize mineral compounds intensively.

Manganese bacteria are a specific group, because they oxidize divalent manganese ions Mn(II) to manganese dioxide (MnO_2) in which manganese is at +4 degree of oxidation. Most commonly, in water distribution networks, in wells and at different stages of drinking water treatment, the following genus of manganese bacteria are found: *Metallogenium personatum, Leptothrix echinata* and *Leptothrix discophora*, which oxidize Mn(II) to Mn(IV). Some iron bacteria are able at the same time to oxidize inorganic manganese salts (e.g. *Sphaerotilus manganifera, Crenothrix polyspora, Crenothrix fusca, Acidithiobacillus ferrooxidans* or genus *Gallionella*). Bacteria able to oxidize manganese originated from the beds of rapid filters belong to (genera): *Leptothrix, Hyphomicrobium, Siderocapsa, Siderocystis, Metallogenium.*

Biological oxidation of manganese occurs at Eh values from +300 to 400 mV and at pH 7.4–7.5. Its chemical oxidation is slow at pH 7.5 and definitely more rapid at pH above 9.0. The presence of high iron concentrations, ammonium and phosphate ions and decreased temperature negatively influences the development of manganese bacteria. Bio-sorption of manganese and its catalytic and biological oxidation occurs on grains of the filter bed which are coated with a layer of manganese oxides. This layer also contains numerous microorganisms present in filtered water. According to Mouchet (1992) the use of biological method allows the elimination of chemical oxidants and a significant increase in filtration rate (to 50 m/h) and mass capacity of the filter bed (to 5 kg/m^2). According to this author, treatment of groundwater by the biological method does not cause technical difficulties and the deposit of iron and manganese oxides (precipitates kept in filtration material) is more susceptible to dewatering. The system of biological groundwater treatment reported by Mouchet (1992) consists of aeration with two-stage filtration and correction of water pH. The correction of water pH is necessary due to different optimal pH values for biological iron and manganese oxidation.

Iron and manganese removal from groundwater

Pressure filtration

Rapid filters are basic parts of many systems of groundwater treatment. They significantly influence the obtained effects and investment and exploitation costs of treatment systems. The active parts of filters are filtration beds in which many complex processes take place. These are physical, physicochemical and chemical processes occurring subsequently or simultaneously. These processes include:

- Transport of a mass (physical processes – molecular diffusion, convection).
- Hydrolysis, ion exchange, oxidation, neutralization (chemical and electrochemical processes).
- Sorption, condensation, flocculation, occlusion, co-precipitation (physicochemical processes).
- Hydraulic transport of microfloccules of deposits in a bed under action of cutting down stresses (hydrodynamic, rheological processes).
- Sedimentation, filtration (physical processes).

The mentioned processes are heterogenic (solid phase-liquid phase). This allows for differentiation of filtration bed zones which act as contact reactors.

In pressure systems, water flowing in the filters contain divalent manganese and iron, mainly in the form Fe(II) (Fe(II) $\gg$ Fe(III)). In beds of pressure filters, the effects of iron and manganese removal depends on catalytic oxidation reactions taking place on the grains of filtration material covered with Fe and Mn oxides. In open systems with reaction chambers, water flowing to the filters contain mostly oxidized iron (Fe(III) $\gg$ Fe(II)) and divalent manganese. Iron is found in the form of microfloccules and floccules. In this case, the flocculation process in the filtration beds determines the effects of iron removal The effects of manganese removal depend on catalytic oxidation of Mn(II).

The process of iron and manganese oxidation in a filtration bed is of autocatalytic character. This means that oxidation products (Fe_2O_3 and MnO_2 oxides) take part in the further oxidation and intensify the process. This fact enables applications of non-reagent technologies (aeration + rapid filtration) for groundwater treatment in a wide range of raw water quality. It is especially important for manganese removal, because with pH characteristic for natural water, manganese is not oxidised with oxygen from the air. Manganese oxidation requires higher redox potential, therefore the zone of filtration bed in which catalytic oxidation of Mn(II) occurs is below the zone of iron oxidation. In the filtration beds, three characteristic zones may be specified:

- Zone I – iron removal by flocculation of microfloccules and floccules of iron oxides, keeping of floccules in pores and channels of the bed.
- Zone II – iron removal by catalytic oxidation and keeping the iron oxides in the form of $Fe_2O_3 \cdot nH_2O$ grain coatings.
- Zone III – manganese removal by catalytic oxidation and keeping manganese oxides in the form of MnO_2 grain coatings.

The scheme of a filtration bed is shown in Figure 10.5. The condition of obtaining the required effects of iron and manganese removal in the process of filtration is dependent on the height of the bed meeting the requirement:

$$H_{bed} \geq \alpha(H(Fe) + H(Mn))$$

Thus, the filtration bed must be of height equal or higher than the sum of the height of zones of iron removal (Fe) and manganese removal (Mn), where α is a coefficient of process reliability and has a value from 1.25 to 1.50.

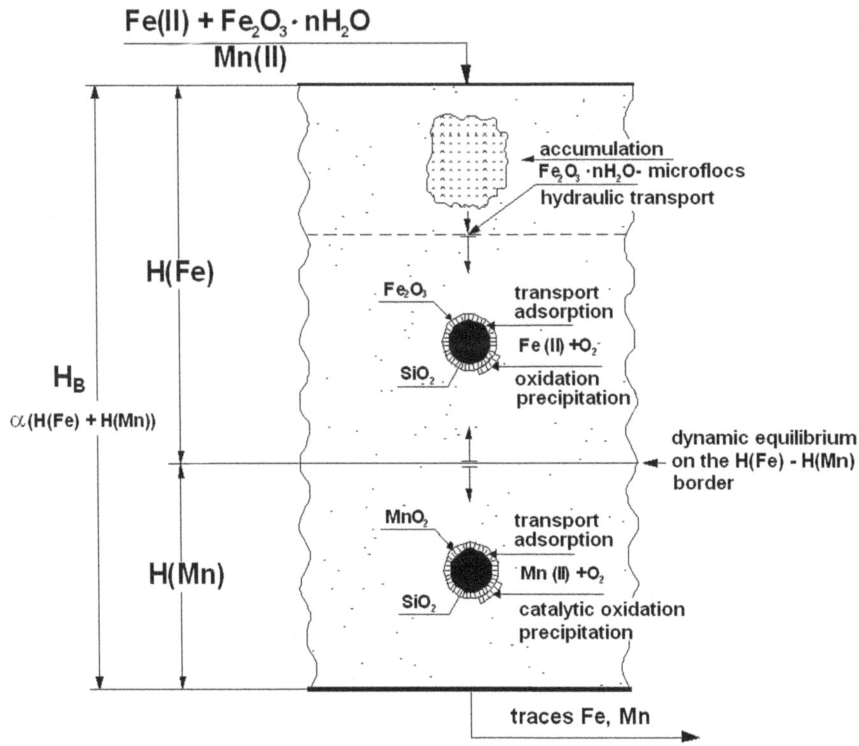

Figure 10.5 Filtration bed schematics.

These zones are more clearly distinguished when optimal conditions of flocculation processes of ferric hydroxide and catalytic oxidation of iron and manganese differ. Together with an increase in the pH and redox potential of water (dosing of chemical oxidants), differences between the bed zones decrease. In case of Mn(II) oxidation and precipitation of MnO_2 before filters, difference between the zones disappears completely.

Catalytic oxidation of Fe(II) and Mn(II) is carried out in filtration beds with oxidative properties, that is, by heterogenic catalysis method. The term "oxidative beds" defines a surface of filtration material on which the processes of catalytic and heterogenic oxidation of iron and manganese proceed. The compound catalysing manganese oxidation is MnO_2 which is a potent oxidizer. It occurs as a component of external layers of filtration material grains. A good example of oxidative bed material is manganese ore, called battery manganese (braunstein). The content of manganese dioxide in manganese ore reaches sometimes over 90%.

Oxidative beds can be also manufactured using silica sand. In this case the beds must be at first activated, that is, coated with Fe and Mn oxides during filter exploitation.

Activation of an inactive bed, for example, quartz or chalcedonite, may be carried out without reagents. The filtration bed is then coated with layers of oxides Fe(III) and Mn(IV) derived from treated water. In order to carry out the autoactivation process, resulting in the coating of filtration bed grains with durable and active coatings of iron and manganese oxides, a number of conditions must be fulfilled. This includes raw water quality, filtration material properties and the autoactivation balance determined by adsorption and desorption of manganese during filtration and backwashing.

Physico-chemical conditions concerning the quality of raw water are as follows:

- Concentration of the oxygen dissolved in water should be equal or higher than oxygen demand resulting from the content of iron Fe(II), manganese Mn(II), ammonium nitrogen and other admixtures of oxygen.
- Concentration of manganese in raw water should be higher than 0.2–0.3 mg/l.
- Lack of reducing agents. For instance the presence of Fe(II) precludes the formation of a manganese removal zone or leads to its deactivation.
- Manganese in raw water should be in the form of mineral compounds determined by a condition: $CODKMnO_4 \leq 3 \div 5\ mgO_2/l$.

Adsorption properties of filtration material are determined by the value of susceptibility of this bed to activation (W.P.):

$$\text{W.P.} = \frac{[\text{Mn}]_A}{[\text{Mn}]_C} > 0$$

Ratio of adsorbed manganese $[\text{Mn}]_A$ to total manganese $[\text{Mn}]_c$ may be expressed by the equation:

$$\frac{[\text{Mn}]_A}{[\text{Mn}]_C} = \frac{[\text{Mn}]_{rw} - [\text{Mn}]_F}{[\text{Mn}]_{rw}}$$

wherein: $[\text{Mn}]_{rw}$ is a manganese content in raw water, and $[\text{Mn}]_F$ is a manganese content in a filtrate. Thus, the ratio of manganese concentrations in the filtrate and manganese in raw water should be below 1:

$$\text{W.P.} = \frac{[\text{Mn}]_F}{[\text{Mn}]_{rw}} < 1$$

The balance of adsorption and desorption of manganese should indicate that during activation of the bed, the mass of manganese kept during filtration cycle is higher than the mass of manganese removed from the filter during backwash.

The mass of manganese kept in the bed during filtration cycle is equal to the product of manganese concentration difference between an inlet and an outlet from the filter, filtration rate and duration of filtration cycle.

The mass of manganese removed from the bed during backwash is a product of mean manganese concentration in backwash water, backwash intensity, and time of backwashing with water.

During the stable work of the filter the amount of manganese kept in the filter is higher or equal to the amount of manganese that is released from the filtration bed during back washing. Deactivation of the bed occurs when more manganese is removed from the filter during backwash than is kept in the filter during filtration cycle. Then, a decrease of catalytic coating of bed grains occurs.

If auto-activation conditions are not fulfilled, the filtration beds can be activated with the use of dosing of potent oxidizers into raw water, most commonly, potassium permanganate. During dosing of $KMnO_4$ into the solution, grains are coated with manganese dioxide derived from $KMnO_4$ decomposition and from oxidation of Mn(II) contained in treated water. Catalytic coatings forming during chlorine dosing contain manganese derived from treated water. Difficulties in the process of auto-activation of the beds, especially in small water treatment plants, result in increased popularity of natural manganese ore sands or manufacturing activated oxidative beds.

Efficiency of filtration through oxidative beds depends upon parameters of internal structure of catalytic coatings that is, pore surface (PW) and pore volume (OP) as well as percentage of manganese part manganese dioxide in coatings $[MnO_2]$. Values of parameters defining general criterion of high catalytic efficiency of manganese removal from groundwater in oxidative beds are (Jeż-Walkowiak 2000):

$$OP \geq OP_{min} = 0.03\ cm^3/g$$

$$PW\text{-}BET \geq PW\text{-}BET_{min} = 9.0\ m^2/g$$

$$[MnO_2] \geq [MnO_2]_{min} = 45\%$$

Significant differences in structures of precipitating $Fe_2O_3 \cdot nH_2O$ can be found as a result of oxidation, hydrolysis and flocculation (Grochulska-Segal & Sozanski 1984). When precipitation takes place in the reaction chambers before rapid filters, iron oxides precipitate in the form of floccules of lower packing solid phase in the structure. These large size floccules show high susceptibility to de-dispergation under influence of hydrodynamic forces. Ferric oxides formed during processes occurring in the oxidative filtration beds are in the form of relatively stable coating of $Fe_2O_3 \cdot nH_2O$ of substantial packing of solid phase. They show higher resistance to action of hydrodynamic forces. Thus, the typical characteristic of the structure of ferric oxides formed as a result of oxidation by a method of heterogenic catalysis in the oxidative beds is its higher hydrodynamic stability in relation to the structure of oxides precipitated as a result of oxidation under homogeneous conditions. The coatings of $MnO_2 \cdot nH_2O$ formed during manganese removal by heterogenic catalysis in oxidative beds are characterized by high hydrodynamic stability. This allows the use of much higher filtration rates than in the process of filtration of water containing oxidized iron in forms of microfloccules and floccules.

The above discussion concerns iron and manganese found in water in the form of mineral compounds. If water shows increased colour and turbidity, iron and manganese are usually found in organic complexes. In these cases water treatment should be expanded into chemical oxidation processes, contact filtration or combined coagulation and sedimentation in sludge blanked clarifiers. Then, the effects of oxidation, coagulation and sedimentation determine the effects of treatment.

Processes using rapid gravity filtration

The iron floc produced when anoxic groundwater is aerated (or treated with chemical oxidants) has a positive surface charge, thus the flocs repel each other and are therefore difficult to coagulate, settle and filter. Physico-chemical treatment rates can therefore be low, from 1 to 2 m/h for settlement and 4 to 5 m/h for filtration to obtain optimum treatment. These rates can be enhanced by the presence of naturally occurring negatively charged particles which neutralise the repulsive forces. However, these are rarely present in groundwaters and materials such as clays must be added or in hard waters a degree of softening used, usually by the addition of lime, to produce calcium carbonate particles. Filtration rates can be raised by autocatalytic action in the filters as described under Pressure Filtration.

Filter capacity is also limited and at iron concentrations above 2 mg/l filter backwash frequencies can become unacceptably short and manganese removal may be compromised. In these cases the load to the filters should be reduced using a settlement stage or microbiological filter before the rapid gravity filters.

The choice of a treatment method will depend on the water quality which for many groundwaters can be categorised as:

- Iron only.
- Iron and manganese.
- Iron, manganese and ammonia (and hydrogen sulphide).

Iron only

Iron at concentrations below 2 mg/l can be removed by aeration and rapid gravity filtration. For the optimum performance a filter bed depth of 1 m should be used consisting of 90% of 0.65 mm (average grain size, 16/30 mesh) sand, and 10% of a suitable grain size of granular manganese dioxide (sized such as to stay mixed with the sand after backwashing, this will vary with the source of the manganese dioxide). Filtration rates are conservative (4 to 6 m/h). Good performance can also be achieved with other filter media such as greensand and treated sands (see Pressure Filtration).

If the iron concentration is above 2 mg/l then a settlement (2 m/h) or microbiological filter stage (50 m/h) may need to be used prior to the filters.

If organic complexation of the iron is a problem, then chemical oxidants may need to be used after aeration.

Iron and manganese

The same treatments as for the removal of iron only may be used for the removal of both iron and manganese with the addition of chemical oxidation after aeration.

Iron, manganese and ammonia

Again the same treatment as for the removal of iron and manganese may be used with chlorine as the chemical oxidant. In order to achieve breakpoint chlorination of the ammonia a chlorine contact tank of 10 to 15 min contact time is used. This is normally placed after aeration in order to facilitate contact time in the tank and filters towards disinfection requirements. This sequence will also deal with hydrogen sulphide but care must be taken to properly vent the toxic hydrogen sulphide gas stripped during aeration.

Microbiological filtration

Many groundwaters are suitable for the employment of microbiological removal of iron and manganese (see Microbiological Oxidation). The processes are high rate, require no chemical oxidants and can therefore be highly economic. The iron and manganese must be in their reduced states.

Iron only

The water is aerated to 8% dissolved oxygen saturation and filtered. A typical filter is a 2 m deep bed of coarse sand (1 to 1.5 mm average diameter, 8/16 mesh). This can be operated at up to 50 m/h and will treat iron concentrations above 2 mg/l. Backwashing is preferably using simultaneous air and water. Treatment establishes within a few days. Other filter configurations and backwash regimes have also been successful. The water must be aerated before distribution.

Iron manganese and ammonia

Unfortunately microbiological manganese removal requires at least 60% dissolved oxygen saturation and therefore after iron removal as above the water must be aerated before a second filtration stage. Treatment takes about 6 weeks to establish. Treatment rates are quoted as up to 40 m/h but practically, ammonia is also often microbiologically oxidised in the manganese removal filter and is the rate – determining process.

For Northern European groundwater temperatures (11°C) this means about a 15 min contact time and a process rate of about 12 m/h, dependent on the filter design (see Figure 10.6).

Sequestration

In the processes of water treatment, corrosion inhibitors are used more and more often. They belong to three main groups: phosphates, silicates and organic inhibitors.

Phosphate corrosion inhibitors (orthophosphates, polyphosphates and blends of ortho- and polyphosphates) have the longest tradition of use – they were used for the first time in 1900 (McNeil & Edwards 2000) and their use is now common.

One of many effects using corrosion inhibitors in water treatment process is the stabilization of water quality. Iron Fe(II) and manganese Mn(II) ions derived from intakes of groundwater and penetration into water as a result of processes of secondary water contamination in the water distribution system, are masked.

In ground and well water, iron and manganese are present in the form of insoluble bivalent ions Fe(II) and Mn(II). As a result of contact with oxidant (oxygen, disinfectants), these ions are oxidized to the trivalent ions Fe(III) and Mn(III); these forms are insoluble and coloured. Insoluble iron salts are the cause of "red water", which can cause reddish-brown stain on washing kitchen utensils and is related to many claims from consumers concerning organoleptic properties of water. Manganese compounds are subject to similar transformation leading to formation of "black water", what may result in formation of black-brown stains on contacted surfaces. These effects may be minimised by maintaining the waterworks output concentrations at 0.05 mg/l iron and 0.02 mg/l manganese, however it is not likely to be completely eliminated.

In order to retain iron and manganese in solution, polyphosphates are used (DHS 1992; MWH 2005). Use of 2–4 ppm of polyphosphates in the form of sodium hexametaphosphate, sodium tripolyphosphate or tetra-sodium pyrophosphate before chlorination results in forming of colourless phosphate complexes of metals and eliminates formation of their insoluble compounds (American Chemistry Council, 2004). Polyphosphates are not so effective in control of colour, when iron is already present in the form of

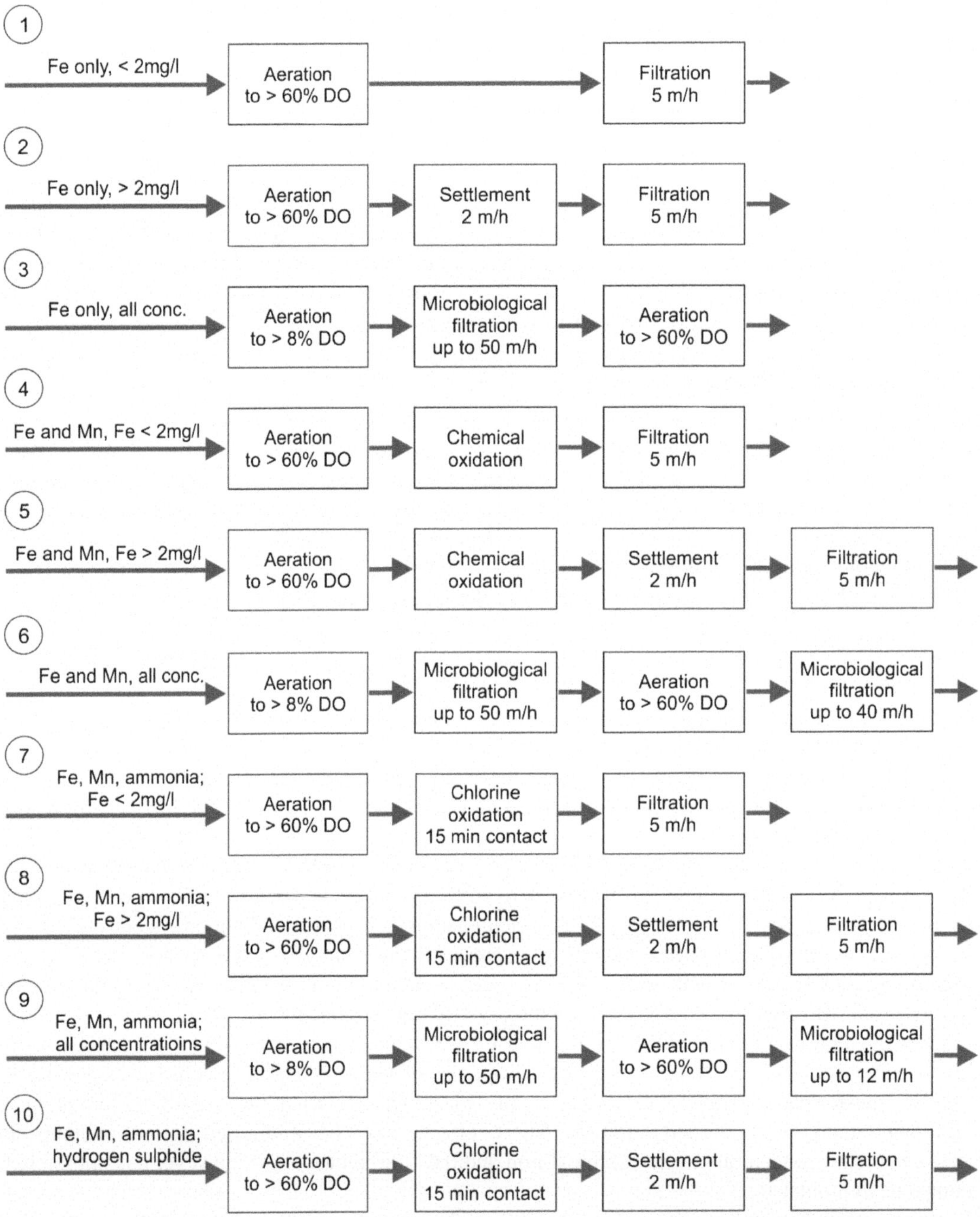

Figure 10.6 Summary block diagrams for popular iron and manganese removal technologies.

Fe(III). Therefore, it is most valuable to add polyphosphates before chlorination or aeration; these processes oxidize Fe(II) to Fe(III).

Doses of polyphosphates required in the process of masking depend, among other things, on water hardness. In the case of lack of hardness, it was observed that a dosage of about 1 mg of PO_4^{2-}/l of polyphosphates was required to mask 2 mgFe/l. In presence of carbonate hardness at the level of 100 $mgCaCO_3$/l, 5 mg of PO_4^{2-} was required (MWH 2005).

The complexes formed are stable in the presence of chlorine at the level found in chlorinated drinking water. At the same time, no reaction is observed which would reduce the level or effectiveness of action of both chlorine and polyphosphates. Iron and manganese masked in the form of colourless complexes before chlorination, will also remain colourless after the process of water disinfection (DHS 1992; ACC 2004).

Sodium silicate has also been successfully used to sequester Iron in the reduced form. This is achieved by the addition of 1 mg/l of sodium silicate and the complex is stable to boiling.

It must be remembered that sequestration does not remove the iron and manganese and therefore statutory upper concentration limits may be exceeded.

Other methods

Ion exchange may also be used for removing low concentration (<0,5 mg/l) of Fe(II) and Mn(II) from groundwater. The majority of ion exchange applications for iron and manganese removal are limited to treatment of industrial water. Usually a strong-acid cation exchange resin in the sodium form is used (MWH 2005).

Nanofiltration membranes are very effective for the removal of soluble iron and manganese. However even a small amount of oxidized iron and manganese can foul membranes and cause a decrease in their effectiveness.

Iron and manganese removal from groundwater in the aquifer

The traditional methods of iron and manganese removal require usually expensive equipment and appropriate maintenance in order to ensure proper functioning. It is particularly important to strictly follow technological procedures of flushing filters. Back wash water produced as a result of this process poses an environmental problem.

An alternative and cheaper way of removing iron and manganese from water may be the method of water treatment in the aquifer. This method does not only enable the reduction of investment and exploitation costs when compared with traditional treatment methods, but it also protects the well screen and other devices used for the intake and distribution of water from the process of clogging with iron compounds. The method of treatment in the aquifer was initially used in the 1960's in Finland and it was then developed by a Swedish company, Vyredox (Hallberg & Martinell 1976). The method has been researched in many other countries all over the world. It involves the artificial creation of an oxidation zone around the well screen using injections of water with oxidizer, most frequently oxygen dissolved in water (aerated and degassed water).

In the course of works on the application of the method in different countries, various technical and technological versions of its use have been invented. The versions can be divided in terms of the conditions of well-field exploitation into:

(1) Periodic injection of water oxidized in the aquifer, with the well not working during the injection,
(2) Continuous injection, with the well working during the injection.

The application of version 1 involves periodic injection of aerated and degassed water to the aquifer around the well (Figure 10.7). The efficiency and time of injection should provide conditions for the formation of properly sized active zone in the aquifer, which will enable the oxidation of iron and manganese compounds and their filtration in the water-bearing deposit to ensure the flow of treated water into the well. With the injection yield equal to the exploitation yield of the well, the injection period of 1 day is usually sufficient, after which the well can be exploited again.

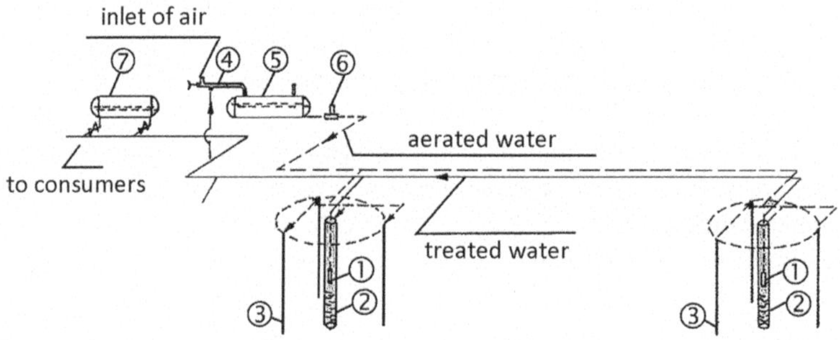

Figure 10.7. Installation for the removal of iron and manganese from groundwater in aquifer – the version of periodic well-field operation. 1 – submersible pump, 2 – well, 3 – injection hole, 4 – ejector, 5 – water degassing tank, 6 – oxidized water pump, 7 – hydrophore

In order to inject the aerated water into the aquifer, different solutions can be used:

- Boreholes around the well which allow the reduction of the number of injection boreholes to three.
- The well and boreholes around the well.
- The well and one injection borehole.
- The well itself.

Version (2) mentioned above, developed by Vyredox (Braester & Martinell 1987), involves a number of injection holes placed around the well at the appropriate distance (Figure 10.8). Each borehole must be equipped with a pump. Aerated and degassed water is injected into one of the holes, and the two nearest boreholes are pumping at that time. By steering the injection and exploitation respectively using a microprocessor, appropriate distribution of aerated water is obtained in the whole area between the injection boreholes and the continuously working well. Due to the necessity to make many injection boreholes and its complicated system of steering the injection, the above solution is not commonly used despite its many advantages when compared with version (1).

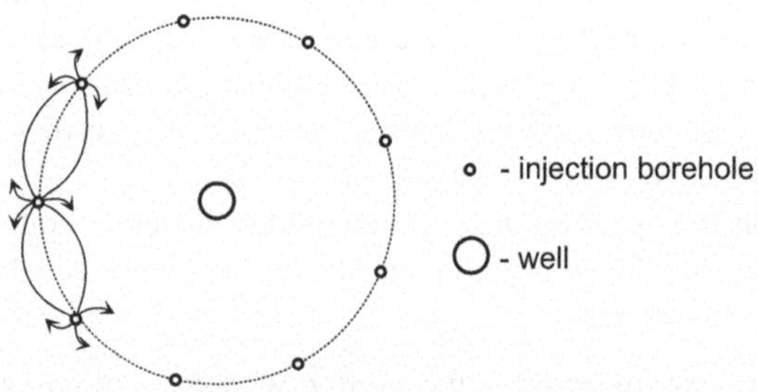

Figure 10.8 Diagram of system of water injection to aquifer – continuous well operation

The method of water treatment in the aquifer uses the catalytic influence of ferruginous and manganese bacteria on the development of iron and manganese oxidation processes. In this connection, the efficiency coefficient of water treatment is defined by the proportion of treated water to water used for injections increases in the first months of using the method, along with the development of bacteria cultures. The increase in efficiency is also related to:

- The increase in adsorption properties of the water-bearing deposits in relation to oxygen
- Catalytic influence on the processes of oxidation of the newly precipitated iron and manganese hydroxides.

After several months of using the method, the efficiency coefficient stabilizes, and its value depends on the hydrogeochemical conditions in the water-bearing layer and it may vary within the wide range from 3 to 50. The application of the method also allows for obtaining significant increase in water quality in terms of such parameters as colour, oxidisability and ammonium nitrogen (Table 10.4). It should be underlined that iron and manganese compounds precipitated from water are deposited in large parts of aquifer, which allows for the effective working of the well in the perspective of tens or even hundreds of years.

Table 10.4 Examples of results of water treatment in aquifer.

Indicator	Wroniawy well-field (Poland)		Grimsäs well-field (Sweden)	
	Raw water	Treated water	Raw water	Treated water
Colour [mg Pt/l]	15	5	85	<5
Oxidisability [mgO$_2$/l]	2.5	2.4	4	2
Reaction [pH]	7.0	7.3	6.6	–*
Total iron [mg/l]	1.1	–	3.9	<0.005
Manganese [mg/l]	0.2	0.03	0.07	<0.005
Ammonium nitrogen [mg/l]	0.3	0.09	0.8	<0.1

–*not determined.

Both theoretical calculations and observations from the functioning well-fields indicate this. Observations also indicate that, with the proper application of the method, no ferruginous deposits on well screen are observed, which is the most common problem experienced during the traditional exploitation of wells. The described method is successfully used in many countries worldwide.

It is estimated that this concerns about 1000 wells in 20 countries. In Poland it has been successfully applied for over 20 years at a large well-field supplying the town of Wolsztyn and at several smaller well-fields supplying villages (Górski 2011).

KEY POINTS:

(1) Sediments and biofilm decrease the pipe diameter leading to an increase of hydraulic losses and energy costs of water distribution.
(2) Bacterial and chemical corrosion significantly reduces the degree of reliability of a water distribution network causing an increase of cost of the repairs.

(3) Iron and manganese removal is based on the precipitation of hydrated iron oxides $Fe_2O_3 \cdot nH_2O$ and manganese oxides $MnO_2 \cdot mH_2O$, which are practically insoluble and thus easy to remove from water by filtration. The classic technology of water treatment is based on transformation of Fe(II) and Mn(II) compounds to the form of insoluble oxides of Fe(III) and Mn(IV).

(4) An alternative and cheaper way of removing iron and manganese from water may be the method of water treatment in the aquifer.

(5) Corrosion inhibitors can be used to achieve the stabilization of water quality.

Chapter 11

Economics of iron and manganese control

11.1 MAIN ISSUES

The economics of iron and manganese in water can be considered in various aspects, including:

- Costs of new treatment plant construction.
- Upgrading (retrofitting) existing plant for more efficient iron/manganese removal.
- Costs of corrosion control in water mains.
- Costs of pipe replacement.
- Cost of iron/manganese removal in local systems.
- Indirect benefits such as the usage of iron-rich sludge from wastewater treatment.

11.2 COST OF WATER TREATMENT PLANT FOR IRON AND/OR MANGANESE REMOVAL – NEW PLANT

Large size water treatment plants

The water treatment plant flow scheme is similar for iron removal only (IR) and iron and manganese removing (IR/MR) plants. The cost of iron removal in mid-scale water treatment plants (WTP, depends mainly on WTP capacity expressed as a design flow.

New plants

Cost increases from IR to IR/MR plant are caused by the use of specific filtration media with manganese oxidation possibilities (e.g. Greensand©, Birm©/Filox™ etc.) and supporting equipment ($KMnO_4$ dosing) if necessary. The volume of filtration media depends on hydraulic loading during filtration (indirectly, it depends on raw water quality).

The unit price of this specialized media may vary depending on the type of material and the supplier's terms of trade. Usually this price is 3 to 7 times higher than conventional filtration sand bed.

The unit price of a manganese removing filtration bed ranges from approximately 700 to 3000 €/m^3 (i.e. The highest unit price is a catalogue price for small capacity filter replacement set of high quality material (US$ 144 per cubic foot). Assuming a hydraulic loading of 5 m^3/m^2 · hour and 1 meter filtration bed depth, the volume of a specialized filtration bed for 10,000 m^3/day WTP can be estimated as 83 cubic meters which will cost (assuming bulk purchase) between € 70,000 and € 90,000.

As the total volume of filtration bed depends on the hydraulic loading of filtration, this parameter should be determined accurately due to its impact on the cost of filtration media.

McGivney and Kawamura (2008) estimated costs for large water treatments plants of capacity varying from 10 to 100 mgd (US) which corresponds to approx. 3800 to 38000 cu.m. per day (m^3/day), showing that total investment cost tend to increase slower than capacity of the iron and manganese removing plant capacity (McGivney & Kawamura 2008). Relationship between plant's capacity and costs were statistically generalized as a power function: $y = 10.636x^{0.5584}$ for total capital costs and $y = 7.8783x^{0.5584}$ for total construction costs, where x = plant capacity in mgd (US) and y = costs in millions of US Dollars. The same authors established the formula for Operation and Maintenance (O&M) costs as y = 7.8783 Table 11.1 shows example calculations for selected plants' capacities based on these equations.

Table 11.1 Cost simulation for iron and manganese removal WTPs.

WTP capacity		**Total construction costs**		**Total investment costs**	
mgd (US)	**m^3/day**	**Million USD**	**Milion Euro**	**Million USD**	**Milion Euro**
10	37,850	28.5	22.0	38.5	29.6
13,2	50,000	33.3	25.7	45.0	34.6
20	75,700	42.0	32.0	56.7	43.6
26,4	100,000	49.0	38.8	66.2	51.0
50	189,250	70.0	54.0	94.5	72.8
52,8	200,000	72.2	55.6	97.5	75.1
79,3	300,000	90.6	69.8	122.2	94.1
100	378,500	103.1	79.5	139.2	107.2

Cost data were recalculated to Euro with exchange rate 1 €=1298 as for 1st January 2013.
Source: Price level (2008).

Mid-size water treatment plants

Economics of water treatment plants of mid-size capacity (up to 3 mgd/approx. 10000 m^3/day) removing iron and manganese, depends on the WTP's capacity as well as on water alkalinity. Water of high alkalinity (over 250 mg $CaCO_3$/l) requires low efficient aeration usually performed by a pressure vessel aerator supplied with compressed air from a compressor.

High alkalinity water

Unit treatment costs for IR/MR systems tended to decrease with increasing water treatment plant capacity. Table 11.2 summarizes unit treatment and total capital costs calculated on a basis of the US EPA report on cost of arsenic removal, which uses the same technology as iron and manganese removal (US EPA 2011) and data obtained from Polish WTPs (Rybicki 2010).

Unit treatment costs for IR/MR systems in the year 2011 tended to decrease with increasing water treatment plant capacity. The average unit treatment cost for mid-size plants in Poland are summarized in Table 11.2.

Examples of total capital costs for the newly built plant are presented in Table 11.3. Data obtained from US EPA Report (US EPA 2011) show maximum and maximum values from investigations. These data do not include cost of building and/or land purchase, legal costs and so economics of these examples is more 'optimistic' than by McGivney and Kawamura.

Table 11.2 Unit treatment cost example for mid-size iron and manganese removal WTPs.

WTP capacity		Unit treatment costs	
mgd(US)	m^3/day	USD/1000 gal	Euro/m^3
0.26	1000	2.05	0.42
0.66	2500	1.46	0.30
2.64	10,000	0.59	0.12

Source: Price level (2010).

Table 11.3 Total capital cost example for small to mid-size iron and manganese removal WTPs.

WTP capacity		Total capital cost		References	Comment
mgd(US)	m^3/day	Million USD	Milion Euro		
0.26	1000	0.26	0.20	Rybicki (2010)	
0.61	2500	0.45	0.35	Rybicki (2010)	
2.64	10,000	0.75	0.58	Rybicki (2010)	
25	95	0.055	0.042	US EPA (2011)	Non-transient, non-community system, min
60	227	0.128	0.099	US EPA (2011)	Non-transient, non-community system, max
20	76	0.065	0.050	US EPA (2011)	Small community system, min
96	364	0.162	0.125	US EPA (2011)	Small community system, max
140	530	0.220	0.169	US EPA (2011)	Mid-size community system, min
770	2900	0.43	0.331	US EPA (2011)	Mid-size community system, max

The structure of the investment cost was (on average): 60% for equipment purchase, 15% for site engineering and 25% for installation.

Low alkalinity water

Water of lower alkalinity (below 250 mg $CaCO_3$/l) requires high efficient aeration usually performed by an open type aerator (spring or cascade) followed by a reaction chamber with a hydraulic retention time (HRT) between 1 and 2 hours. A unit treatment cost for low alkalinity IR/MR systems were calculated based on the same data as for WTP working with high alkalinity water (see Table 11.4).

Table 11.4 Total capital cost and unit treatment example for small to mid-size iron and manganese removal WTPs treating water of low alkalinity.

WTP capacity		Total capital cost		Unit treatment cost	
mgd (US)	m^3/day	Million USD	Milion Euro	US$/1000 gal	€/m^3
0.26	1000	0.33	0.25	2.52–3.15	0.52–0.65
0.61	2500	0.48	0.37	1.65–2.28	0.34–0.47
2.64	10,000	0.71	0.55	0.68–0.87	0.14–0.18

Additional costs taken into consideration were additional investment costs (Rybicki 2010):

- Construction of a 1.5 hrs HRT reaction chamber supported by a cascade aerator.
- An additional pumping system to lift water from the chamber to filtration unit.

Normalized capital costs were estimated to be € 250/m^3/day; € 145/m^3/day and € 55/m^3/day respectively. The structure of investment cost was similar to units for high alkaline water treatment units (US EPA 2011; Rybicki 2010).

Low capacity systems

Low capacity treatment systems are usually operated with average daily production lower than designed value due to diurnal variation of water consumption. It may lead to much higher unit costs when calculated per cubic meter of water being consumed: it is 5 times higher for 1000 m^3/day plant and twice higher for the largest plants tested. Considering the difference between these two ways of estimating unit treatment costs is important in decision-making procedures. Results obtained by the US EPA for WTPs of similar capacities were obtained for systems addressed for arsenic removal; however both treatment technology and facilities applied were similar to Fe removal plants, showed similar results (US EPA 2011).

For example for non-transient non-community systems operation and maintenance unit costs were established as: 0.45 US$/m^3 for the system of 225 m^3/day capacity and 0.41 US$/m^3 for the system of 540 m^3/day capacity respectively (unit costs were recalculated from US to metric units). Operation and maintenance unit costs for community water systems of 270 m^3/day capacity was estimated 0.31 US$/m^3 (US EPA 2011). These data should be considered as illustrative and exact cost calculations for specific locations may differ significantly.

11.3 COST OF RETROFITTING WATER TREATMENT PLANT FOR IRON AND/OR MANGANESE REMOVAL

Retrofitting existing plants for efficient iron and manganese removal is usually done with the use of existing filtration vessels, but with the exchange of the filters' under-drains, replacement of the filtration media and changes to the chemical dosing unit when necessary. In recent years retrofitting has very often included pipe installations made of stainless steel, which prevents against corrosion during the treatment process.

The cost of improving treatment quality at the IR plant can be estimated (price level for the year 2011) to be around € 70,000 to € 120,000 (90,000 to 150,000 US$) for a 1000 m^3/day plant, whilst for a 10,000 m^3/day WTP plant the retrofitting cost can be calculated to be € 120,000 and € 280,000 (155,000 to 360,000 US$ respectively (Rybicki 2010; US EPA 2011).

The cost of improvement of treatment quality from IR to MR can vary significantly depending on the choice of filtration media.

11.4 NON-COMMUNITY FACILITIES FOR IRON REMOVAL

Small IR and MR systems (i.e. below 2 cubic meters per hour of design capacity) are characterized by higher unit costs.

Costs for the purchase of complete iron removal equipment (assuming no building is necessary) in there year 2011 was estimated as ranging from 550 to 700 US$ (i.e. between € 300 and € 550) per 1 m^3/hour design capacity (US EPA 2011).

The unit price tends to decrease with increase of capacity of the facility. Unit costs cannot be exactly estimated for these solutions due to the high daily inequality of water consumption. Prices for manganese removal treatment sets are usually 20% higher. One should note that media replacement costs are very high in these cases.

11.5 PIPE REPLACEMENT COSTS

Houses

Pipe replacement costs for domestic pipe-work may range from € 300 to € 2500 per house, depending on the total pipe length and installation mode (open or covered by plaster).

Water mains replacement and rehabilitation

Replacement of pipelines is justified mainly by deterioration of water quality during its transition from a source to customers, and corrosion is one of main causes of this deterioration. In year 2011 the AWWA total capital need for replacement of transmission lines and distribution mains as 58.1 $ per foot (i.e. approx. 147 € per meter) including valves, meters and related equipment (AWWA 2001). The cost of water mains replacement may vary especially in large diameter pipe replacement, as the cost of pipe purchase is only approximately 20–40% of total costs (due to costs of road renovation necessary during replacement works). Conventional replacement of pipelines that is, excavation of existing lines and their replacement using new tubes is difficult to perform in densely built-up area. For example intensive road traffic which is often responsible for pipe failures may hinder such a replacement (LeGat & Eiseneis, 2000; Seagrov *et al.* 1999). That is why pipe rehabilitation is being recognized as a technically and economically feasible alternative for pipe replacement, where pipelines are being repaired on-site. Such trenchless technologies do not require large construction/excavation works which usually affect city traffic. The most commonly applied methods of mains' rehabilitation are (Saegrov *et al.* 1999; AWWA 2001; Piechurski & Dzikowski 2012):

- Slip-lining, which usually involves lining with polyethylene pipes in existing pipelines. Materials used for slip lining include high density polyethylene, polypropylene, fiberglass-reinforced polyesters, reinforced thermosetting resins, and PVC. Of those, high density polyethylene (HDPE) is most commonly used because it is flexible, resistant enough to withstand the tensile and compressive forces encountered through installation, and does not corrode;
- Pipe bursting or burst-lining, methods which involves the replacement of an existing pipe with a new PE pipe of the same (or even larger diameter). Pipe bursting is the only trenchless technology that allows replacement of an existing, deteriorated pipeline with one of (one or two sizes) larger diameter. However the upper limit for the diameter of replacement pipe is about 30 inches. One disadvantage to pipe bursting is that services must be excavated and connected.
- Swage-lining, when a continuous polyethylene pipe of a slightly larger diameter than the existing pipeline, by using hydraulic pressure the new pipe is forced through a die which temporarily decreases its diameter (by 7 to 15%) such that it will fit into the old pipe, and the new pipeline is pulled into place inside the existing pipe. Swage-lining is usually recognized as appropriate where the existing pipe is relatively structurally intact (Wróbel *et al.* 2004).
- Cementing – cement mortar has good anti-corrosive properties for use in water lines, pipes to be replaced are cleaned then a machine sprays a mortar to give a consistent lining thickness on the

pipe wall. This method however should not be applied if low-alkalinity water is to be carried by the pipeline. The total alkalinity of the water should be over 55 mg/l $CaCO_3$ (1.1 meq/l) and pH lower than 9.5 otherwise problems related to calcium compound dissolution and aluminium release to water (from cement) may arise Dąbrowski (2011) after long term investigation of this problem, recommends that for several month after this type of rehabilitation, aluminium and chromium concentration in water must be monitored.

- Spray lining involves cleaning the pipe to remove existing corrosion build-up and then spraying a thin epoxy or PU coating on the inner pipe wall. Sprayed polymeric compounds are having increasing success and achieving rapid cure times. Polymeric lining materials have been used in the United Kingdom for several years and have started to see use in the United States. Around 6000 miles of water mains have been lined with these materials, mostly in the United Kingdom and Canada. In the water sector, the emphasis is shifting towards use of sprayed coatings as semi-structural and structural spray linings (US EPA 2009).

Some pipe rehabilitation methods, like short-lining cannot be recommended for water mains even if they are successfully applied for sewage pipe rehabilitation due to problems with large number of connections between rehabilitation pipe sections.

Cost analysis of pipe rehabilitation for price level 2012 in Poland (Piechurski & Dzikowski 2012) with auxiliary works compared with replacement on conventional way are presented in table calculated for 100 meter of pipeline 150 mm diameter.

In areas of low density of urbanization, the unit metric cost of new pipeline is in fact lower than the pipe rehabilitation cost, however in cases of higher pipe dimension and in densely built areas advantages of rehabilitation over building new pipelines may occur. The cost of road-works is a deciding factor in the decision-making process. It should also be considered that the expected lifetime of contemporary ductile cast iron pipelines is estimated as 100 years while the lifetime of lining and cementation is expected to be no more than 30 to 50 years.

Experience from the implementation of innovative pipe rehabilitation methods in the United States was summarized by Matthews (2012), showing higher unit costs than presented in Table 11.5, which may be credited both to materials' price increase and higher labour costs in the US compared with Poland. Table 11.6 summarises these data.

Table 11.5 Cost example for 100 meters pipe rehabilitation and replacement.

	Euro per 100 meters	US Dollar per 100 meters
Replacement	21,600	28,000
Sliplining	15,300	19,860
Cracking	14,500	18,820
Pipe bursting	15,300	19,860

Source: Price level (2012).

Pipe rehabilitation as well as replacements is relatively costly and decision-making in this field is of importance to water utilities' finances. A conservative approach based on pipeline age cannot be recommended as pipeline reliability depends on material type and the quality of construction rather than simply on pipe age. The choice of best strategy for rehabilitation/replacement may be based on the statistical analysis of pipe breakage.

Table 11.6 Selected technical and economic data from US pipe rehabilitation.

Method	Short description	Unit cost [US$/linear meter]
Semi-structural spray-on lining	Material is sprayed on the inner pipe wall	543
Structural lining (CIPP – cured-in-place pipe)	Flexible liner being placed into the existing pipeline. Steam or heated water is forced into the liner, pushing the liner tightly against the pipe walls. The heat causes the liner material to cure creating a new pipe within the existing	615

Source: After Matthews (2012).

Numerous authors suggest that exponential and logistic GLM models are suitable for these calculations (LeGat & Eisenbeis 2000; Moglia *et al.* 2010; Tabesh & Saber 2012).

11.6 PIPE REHABILITATION STRATEGIES

Modern systems for prioritization of the sections to be rehabilitated are combining GIS (location data and occurrence of breakage in the past) with reliability calculations, combining structural/hydraulic/quality analysis with risk analysis. GIS-based systems combined with conventional flow/pressure network models create a viable procedure form modelling supporting decision making in routine operations (Burrows *et al.* 2000).

Proper strategy of pipeline rehabilitation in municipal water distribution systems is recognised as one of the major problems for water utilities. Water utilities in recent years have tended to move from 'reactionary strategies' (reaction for water quality deterioration with no or little long term planning) towards pro-active based on achieving long-term effects (Engelhardt *et al.* 2000). No specific strategy addressed towards water quality deterioration with respect only to iron presence in water was found in the literature. Rehabilitation strategies applied by water utilities must consider specific demands: quality of water within the distribution system must meet proper standards also the hydraulic performance must comply with regulatory stipulations. Specific factors such as distribution losses, leakage control, maintenance events and operational reliability should be incorporated in such strategies.

Modelling water distribution system performance is an important tool for water companies to choose a proper strategy. Rehabilitation decision models can be classified into groups, integrating these factors. Engelhardt *et al.* (2000) undertook a review of these models with respect to the type of model, economic efficiency, rehabilitation action, hydraulic, network reliability and water quality change. None of the decision models incorporate all these factors, most of the models being examined were based on four factors.

Recent examples on prioritization models for the rehabilitation of water distribution networks combined with GIS (Tabesh & Saber 2012) showed that break rate, infrastructure population and pressure in water mains have the greatest impact on prioritizing the rehabilitation activities. Models based on maintenance records are popular in supporting technical decision for pipeline maintenance, especially for prediction of failure rate rather than prediction of water quality deterioration, some authors (Le Gat & Eisenbeis 2000) found the Weibull Proportional Hazard Model as suitable for prediction of water pipe failures, these authors confirmed that an universal model seemed unlikely to exist. Attempts to find accurate methods for water quality modelling in the network is important in systems when increases of iron/manganese concentration create operational problems, but literature shows that models validated on chlorine concentration or on hydraulic aspects but not on iron compounds (Setiadi *et al.* 2005: Gancel *et al.* 2006). Most pipe failure models use the so called "bathtub curve" as simplified illustration of the survival function of water mains showing impact of time of operation on a failure rate.

The initial (early) part of the curve on Figure 11.1 represents "infantile failure" resulting from errors in pipe laying and initial stage of operation. With respect to iron contamination it may be as a result of improper cleaning prior to operation or remaining welding materials (in steel made tubes). Usually longest is a period of low failure rate (flat "bottom of the bathtub") followed by an increasing failure rate – illustrated as the right-side "bathtub tail end". In considering iron contamination of water this increase may be caused by pipe corrosion (in steel pipes) or by release of iron from pipe sediments. The length of the "bottom" of the curve can be estimated based on historical data, but it must be recognized that long term, complete data from pipeline operation are necessary to use such a model (Hadzilacos *et al.* 2000; Kleiner & Rajani 2001).

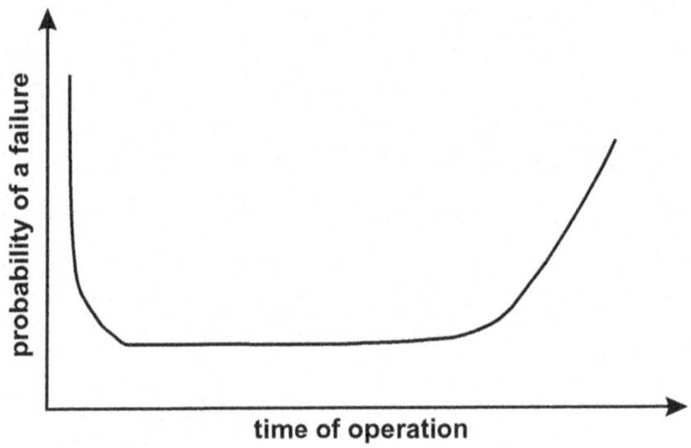

Figure 11.1 Bathtub curve of buried pipeline.

Any changes e.g. partial rehabilitation or replacement make the analysis more complex. Models which not only incorporate failure statistics but also consider physical components of the degradation process make the rehabilitation procedures (ranking) more accurate (Hadzilacos *et al.* 2000).

KEY POINTS:

(1) Unit treatment costs are higher for low alkalinity water when compared to treatment of high alkalinity water, mainly due to post-aeration reaction chamber, which must follow intensive oxidation of iron compounds.
(2) Manganese removal plants (MR) and simultaneous iron and manganese removal plants (IR/MR) generally require higher investment than iron removal plants of similar capacities.
(3) Small IR and MR systems (i.e. below 2 cubic meters per hour of design capacity) are characterized by higher unit costs.
(4) Corrosion of iron-made water mains may enforces replacement or rehabilitation of existing lines to avoid water quality deterioration There is no universal pipe replacement procedure so most utilities apply 'tailor-made' solutions designed for specific water supply systems.

Chapter 12

Developing a strategy

12.1 INTRODUCTION

As should be apparent from earlier chapters, iron and manganese may require control at one or more points throughout the entire water supply chain of "source to tap", either as separate or linked issues. Both metals can impact upon the aesthetic quality of drinking water and the regulatory standards that apply have been set on this basis.

Iron "red water" discolouration is much more common than manganese "black water" discolouration due to the widespread usc (albeit now largely historic) of cast iron for water mains within distribution networks and corrosion. In consequence, large areas of a water supply system can be affected. If source waters are contaminated with either iron or manganese, and if treatment processes are absent or ineffective, then an entire water supply system can be affected, either intermittently or constantly. In all these situations, the level of service provided to consumers can be substantially reduced. In some countries, such as the UK, levels of service are linked to further regulatory controls.

If corrosion of old cast iron water mains is or has been significant, the level of water leakage from the supply system can be as high as 50%, with 25% leakage being fairly common in many older Towns and Cities. Such inefficiencies have two consequences:

(1) More water resource must be abstracted – this could induce "water stress" in the resource if it has limited capacity; there are also environmental consequences of over-abstraction;
(2) The cost of water supply will be higher, particularly in relation to treatment and pumping.

The costs associated with the replacement and rehabilitation of water mains are very high and financial analysis will be justified to ensure that the right balance is achieved between the level of leakage and the cost of its reduction. Water treatment costs, including corrosion control, and the cost burdens of operating a water supply system that is in poor condition should also be considered, such as those involved in rectifying burst mains or flushing to relieve discolouration.

A holistic strategy for controlling iron and manganese in a water supply system must therefore embrace regulatory compliance, levels of service to consumers and asset management; the latter must include a consideration of both capital investment and operating costs.

12.2 RISK ASSESSMENT

Developing a strategy will require risk assessment that is integrated with routine performance measurements. Assessments should be undertaken annually. Table 12.1 lists some of the issues to be considered, recognising the "source to tap" nature of problems with iron and manganese.

Table 12.1 Components of a risk assessment.

Points to consider	Relevance
Source water quality: • confined ground-waters are often contaminated with Fe (and less often with Mn) • rivers that receive mine drainage are often contaminated by metals • manganese can be released from sediments in reservoirs or lakes that stratify in the summer	Ground-waters may require treatment. Surface water treatment may need to be improved.
Use of coagulants containing Fe in surface water treatment. Algal blooms can exacerbate problems.	Fe breakthrough can occur if treatment is not optimised.
The extent of occurrence and location of old cast iron water mains. Problems are more likely if water is stagnant, such as in dead-end mains.	Replacement or rehabilitation may be needed to overcome elevated Fe.
The pH of the water.	If pH is <7, the water can be corrosive resulting in elevated Fe.
Use of corrosion inhibitors.	May need to be optimised (IWA 2012a).
Galvanised iron service pipes and internal plumbing.	May require the use of corrosion inhibitors.
Consumer complaints about discoloured water, particularly if clustered.	May indicate a problem in the distribution network such as a burst water main with associated discolouration. Regular complaints suggest that corrosion of cast-iron water mains is significant.
Microbiological growth within the distribution network is evident from elevated colony counts or coliforms.	Iron release from old cast-iron water mains will be exacerbated by active bio-films within mains and their corrosion tubercles.
Sections of cast-iron water mains, cut out from a repair, show internal tuberculation.	Clear evidence of iron corrosion but needs to be put into perspective (extent and location).
Filter based samplers from water mains collect significant amounts of iron-rust particles or manganese deposits.	Clear evidence of corrosion of the water mains being investigated or the accumulation of manganese.

12.3 COMPONENTS OF A HOLISTIC STRATEGY

Affordability

Consistent with the principles of the Protocol for Water and Health (UN/WHO 2007), drinking water supplies must be affordable to consumers. In this context, it may be necessary to extend regulations beyond the normal water quality criteria that apply in most countries.

Priorities

Any deficiencies with source waters should be resolved before tackling any problems within the distribution network. Once source water problems have been mitigated, the distribution network should be flushed clean, starting at the treatment works and moving outwards to the extremities of the system. Any residual problems from corrosion can then be isolated. The reader should refer to IWA's Best Practice Guide on the Removal of Metals from Drinking Water by Treatment (IWA 2012b).

Engineering standards

Engineering measures must be properly determined by project appraisal before commencement, should be technically sound and meet all relevant engineering standards and codes of practice.

Monitoring

Monitoring must be sufficient for determining the nature and extent of any problems within the system, and for providing a robust basis for corrective actions. Monitoring should continue after mitigation measures have been implemented to demonstrate their success or failure. Consideration should be given to a multiple approach involving several monitoring methods; for example: random sampling of taps, the inspection of water main cut-outs, hydrant flushing, and so on. The reader should refer to IWA's Best Practice Guide for the Sampling of Metals in Drinking Water (IWA 2012c).

Stakeholders and transparency

When engaged in the implementation of mitigation measures, particularly those within the distribution network, Water Utilities should liaise with all relevant stakeholders in a transparent manner and give prior notice to any planned interruptions in supply.

Operational targets and regulatory standards

Consideration should be given to the setting of operational targets that are more stringent than the regulatory standards, as illustrated in Table 12.2. Such an approach should increase the likelihood of achieving compliance.

Table 12.2 Operational water quality targets*.

Target or Standard	Nature of target or standard	Iron [μg/l]	Manganese [μg/l]
Target for treated water	Average	50	10
Target for water at consumers' taps	90th %ile	100	20
Standard at consumers' taps	95th %ile	200	50

*illustrative examples only.

Levels of service

Levels of service can also be defined and will be relevant to the design of mitigation measures. Table 12.3 provides further illustrative examples.

Table 12.3 Levels of service*.

Aspect	Level of service
Burst water mains	Maximum of 3 per annum
Discolouration episodes	Maximum of 3 per annum affecting >10 people No more than 1 per annum affecting >1000 people
Pressure	Maximum 5% population with pressure below 1 bar

*illustrative examples only.

Management principles

The strategy should be reviewed annually and amended as necessary. The responsibilities of all relevant personnel must be clear. All relevant personnel must be competent and given training that is appropriate.

KEY POINTS:

(1) A holistic strategy for controlling iron and manganese in a water supply system must embrace regulatory compliance, levels of service to consumers and asset management;
(2) Developing a strategy will require risk assessment that is integrated with routine performance measurements;
(3) Any deficiencies with source waters should be resolved before tackling any problems within the distribution network;
(4) Engineering measures must be properly determined by project appraisal before commencement;
(5) Monitoring must be sufficient for determining the nature and extent of any problems within the system, and for providing a robust basis for corrective actions;
(6) The strategy should be reviewed annually and amended as necessary.

Appendix 1
Case studies

A1.1 BOGUCIN (PL) – IRON AND MANGANESE REMOVAL FROM A HIGH ALKALINITY WATER, IN THE PRESENCE OF ARSENIC

Country	Poland
Area of case study	Bogucin, Lublin Voivodeship, Eastern Poland
Investigation period	Autumn of 2010
Publication reference	Postawa *et al.* (2011)

Drinking water supplier

Bogucin Water Treatment Plant: established in the 1970s, modernised in 2009.
Capacity of Water Treatment Plant: 1008 m^3/d.
Annual production: 367,920 m^3.

Source and quality of raw water

Groundwater – two wells, 82 meters deep extracting water from Cretaceous-Paleocene deposits.

Problem

High iron manganese and arsenic concentrations in raw water (see Table A1.1.1).

Solution

High alkalinity (above 250 $CaCO_3/l$) allows applying the 3 stages pressure system: pressure aeration and pressure filtration, followed by disinfection.

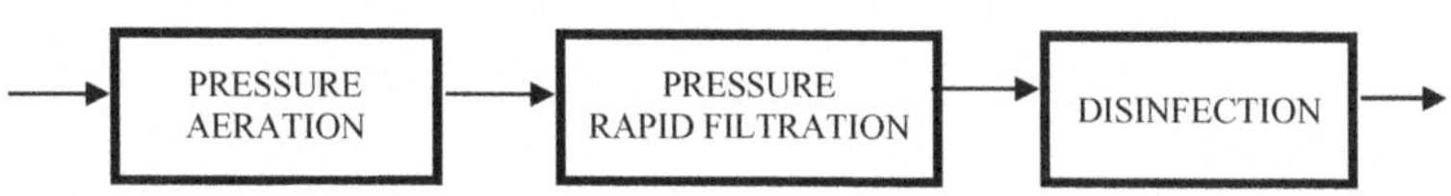

Process parameters:

- Pressure aeration 5,1–10 m^3/h.
- Pressure filtration: 2 filters the diameter of the 1200 mm, 2 filters the diameter of the 1000 mm; filtration rate: 11 m/h; contact time: 6 min; dual media filter bed (Silica sand: H = 0.7 m, d = 0,8–1,4 mm; catalytic mass G1: H = 0,4 m, d = 1,0–3,0 mm).

Table A1.1.1 Bogucin intake. Raw water quality.

Parameter	Value/range
Colour	Colourless
Odour	Odourless
PH	7.0–7.5
Alkalinity [mg $CaCO_3$/l]	340–380
Iron [mg/l]	2.15–3.27
Manganese [mg/l]	0.268–0.665
Arsenic [μg/l]	0.20–0.29

Water-air mixers

Filters

Conclusions

- Three-stage system based on pressure aeration and filtration, followed by disinfection allows achieving high quality drinking water in Bogucin (see Table A1.1.2).
- Relatively high iron concentration in raw water facilitates co-precipitation of iron and arsenic.

Table A1.1.2 Water quality after treatment.

Parameter	Value/range
pH	7.25
Iron [mg/l]	0.002–0.004
Manganese[mg/l]	0.021–0.038
Arsenic [μg/l]	0.000–0.003

REFERENCE

Postawa A., Jeż-Walkowiak J., Pruss A. and Wątor K. (2011). Arsenic in groundwaters near Lublin. Bulletin of Polish Geology Institute. *Hydrogeology*, **445** (12/2), 495–503.

A1.2 SłUPSK (PL) – IRON AND MANGANESE REMOVAL FROM A LOW ALKALINITY WATER

Country	Poland
Area of case study	Słupsk, Pomeranian Voivodeship, Northern Poland
Investigation period	2010
Publication reference	Pruss and Pruss (2010)

Drinking water supplier

Slupsk Water Treatment Plant: modernised in 2007–2009.
Capacity of Water Treatment Plant: 20,000 m^3/d.
Annual production: 7,300,000 m^3.

Source and quality of raw water

Two groundwater sourses: Westerplatte and Globino.
Groundwater from Cretaceous-Paleocene deposits.

Problem

High iron and manganese concentrations in raw water (see Table A1.2.1).

Table A1.2.1 Supsk catchment. Raw water quality.

Parameter	Value/range
Colour	Colourless
Odour	Hydrogen sulphide
pH	7,2–7.6
Alkalinity [mg $CaCO_3$/l]	169–173
Iron [mg/l]	0,81–1.08
Manganese[mg/l]	0.138–0.142

Solution

Due to low alkalinity (below 250 mg $CaCO_3$/l) an open aeration system was applied. Cascade aerators are followed by gravity rapid filters.

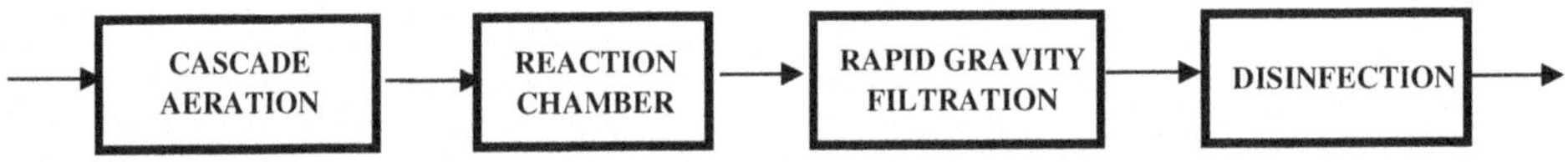

Process parameters:

- 4 cascade aerators (F = 2,2 m^2, O_h = 208 m^3/m^2h).
- 2 reaction chambers (V = 130 m^3; retention time = 25 min).
- 8 gravity rapid filters: Filtration rate 5–7 m/h; F_1= 20 m^2; Contact time: 13–18 min.
- Mono media filter bed: Silica sand d = 0,72–1,25 mm; H = 1,5 m.
- Backwash: air: 3 min, q = 72 m^3/m^2h; water: 10 min, q = 72 m^3/m^2h.

Cascade

Rapid filters

Conclusions

- Cascade aeration and rapid gravity filtration, followed by disinfection allows to eliminate high iron content from water in Słupsk water treatment plant (see Table A1.2.2).
- Open cascade aeration is more effective than pressure aeration when removing iron from low alkalinity water. Aeration introduces oxygen for iron and manganese oxidation and CO_2 removal as well as hydrogen sulphide removal from treated water.

Table A1.2.2 Water quality after treatment.

Parameter	Value
Iron [mg/l]	0.01–0.02
Manganese [mg/l]	0.016–0.031

REFERENCE

Pruss P. and Pruss A. (2010). Water Treatment Plant of The Underground of Technological Research to Start, GWITS. Nov 2010, 16–21.

A1.3 ZIELONA GÓRA (PL) – DETERIORATION OF WATER QUALITY IN TIME

Country	Poland
Area of case study	Zielona Góra Lubuskie Voivodeship, Western Poland
Investigation period	1966-2000
Publication Reference	(Błaszyk and Górski 1981; Górski, 1981).

Drinking water supplier

Water is produced and supplied to the inhabitants by the limited liability company – Spółka z o.o., Zielonogórskie Wodociągi i Kanalizacja".
Number of consumers served: 118,000
The city area: approximately 58 km^2.
Water production: approximately 28,000 m^3/d.

Source and quality of raw water

The city is supplied with water from the Zawada groundwater well-field in the ice-marginal valley of the Oder River and with surface water from the small river Obrzyca, which is the Odra River tributary. Water from both sources is mixed in 50/50% proportion at the water treatment station.

Initially, in the 1960s and in early 1970, the city was supplied only with water from the groundwater well-field. The Zawada well-field was opened in 1966. It consists of 22 wells that extract groundwater from aquifer characterised by the thickness of several to approximately 30 m, locally confined by a layer of muds, 0.5–3 m thick (Figure A1.3.1). The well-field is situated on the flood terrace of the River Oder, protected by a dike in the distance of 3 km from the river.

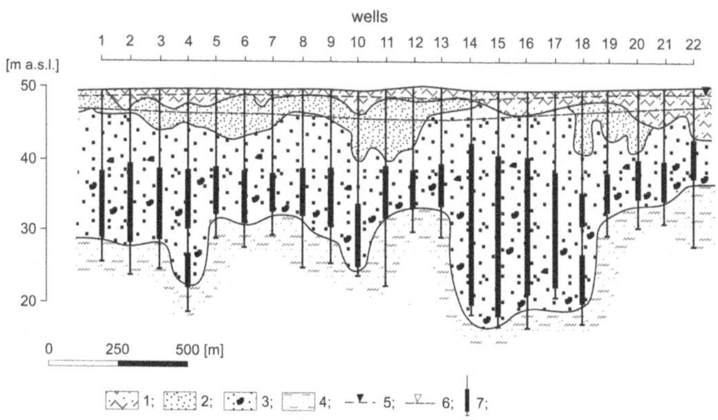

Figure A1.3.1 Zawada well-field; hydrogeological cross-section. 1 – muds, 2 – fine and medium sands, 3 – sands and gravels with pebbles, 4 – silts and clays with sand, 5 – water level before pumping, 6 – water level during pumping, 7 – well screen.

The water from the Obrzyca River is highly polluted in terms of chemistry and microbiology, and it requires treatment in relation to the turbidity, colour, TOC, bacteriology and the content of plankton organisms.

Problem

At the beginning, the quality of groundwater was very good. The iron concentrations did not exceed 1 mg/l, and the manganese concentrations – 0.1 mg/l. During the exploitation, deterioration of water quality in terms of iron, sulphates, manganese and water hardness were observed (see Table A1.3.1). After three years of exploitation, the iron concentrations reached 30 mg/l, the manganese concentrations – 0.80 mg/l and the sulphates concentrations – 350 mg/l (see Figure A1.3.2).

Table A1.3.1 Zawada well-field. Raw water quality.

Parameter	Range
Iron [mg/l]	5–7
Manganese [mg/l]	0.2–0.3

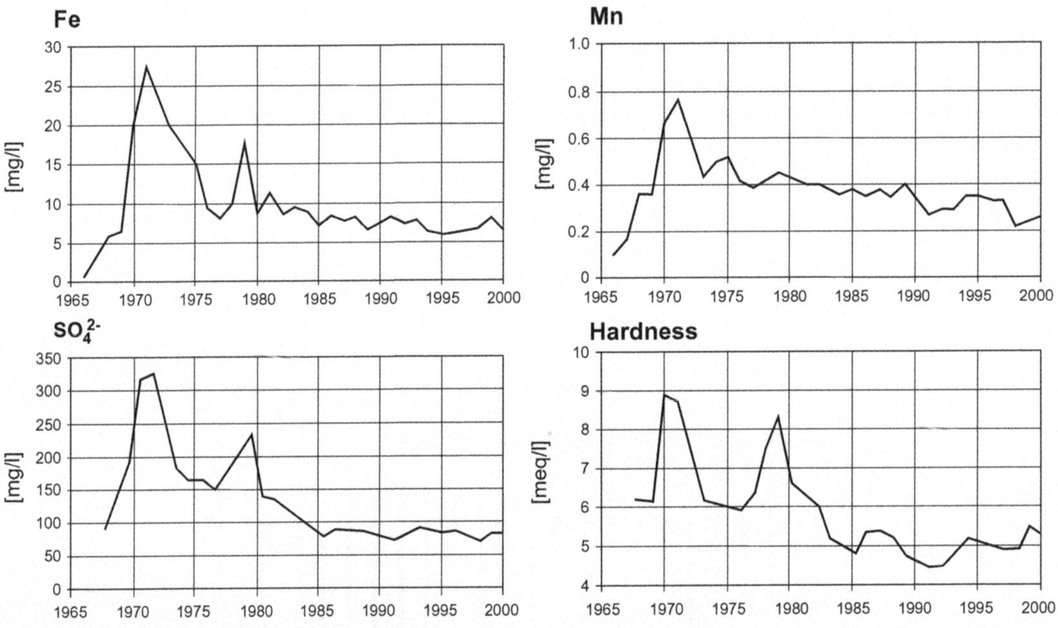

Figure A1.3.2 Changes of selected hydrochemical indicators during the Zawada well-field pumping (hydrochemical data from the collective well).

The detailed geochemical and hydrochemical studies conducted in the early 1970s showed that the cause of the decrease in water quality were the processes of oxidation of iron sulphides. The sulphides were found in the upper part of the water-bearing layer, within fine-grained fluvial formations containing dispersed organic matter. In these formations, naturally confined from the influence of atmospheric conditions, the processes of anaerobic decomposition of organic matter occurred, and S^{2-} ions released in these processes were bonded to iron forming sulphides. Due to exploitation, an aeration zone was formed, where the processes of sulphide oxidation developed, in accordance with the reaction catalysed by sulphur bacteria.

$$2FeS_2 + 7O_2 + 2H_2O\ 2H \xrightarrow{\text{cat.biol.}} 2Fe^{2+} + 4H^+ + SO_4{}^{2-}$$

This caused the release of large amounts of iron and sulphates, and the sulphur acid that was produced caused rapid dissolution of carbonates and iron, as well as manganese compounds occurring in the water-bearing sediments. This process was of a transient nature, but its influence on the chemistry of water obtained from the well-field was noticeable for 19 years. After that the water quality stabilised comparatively. The concentrations of iron and manganese were, however, much higher than before the initial exploitation, when extremely reductive conditions determined the low concentrations of iron, manganese and sulphates.

Solution

The dramatic decrease in the quality of water at the Zawada well-field provoked the decision about the termination of its use and the construction of a surface water scheme. The Zawada well-field still had been exploited until the river water scheme was open in 1976. However, it was not terminated after the opening of the river water scheme due to the technology of using the water from the Zawada well-field with high content of iron to coagulate river water. This technology is still used at present, but because of the significant decrease in the iron content in groundwater, iron sulphate is added as the coagulant.

At present the water from the Obrzyca River runs through grids and microsieves and then is mixed with groundwater in the 50/50% proportion; further stages comprise: coagulation, dosing of chlorine dioxide, filtration on sand filters, disinfection with chlorine dioxide, pH correction using calcined soda (Na_2CO_3), which results in increase of pH from 7.10 to 7.70. Mentioned process allows providing water which meets the quality standards for drinking water.

Conclusions

- Groundwater abstraction, especially in river valley and ice-marginal valley, can cause a groundwater quality determination in the range of iron, manganese and hardness as a result of sulphides and organic matter oxidation.
- In order to eliminate or limit the influence of above phenomenon on the water quality it is necessary to conduct appropriate hydrochemical and geochemical investigation during a well-field designing stage.

REFERENCES

Błaszyk T. and Górski J. (1981). Ground Water Quality Changes During Exploitation, *Ground Water*, **19** (1), 28–33.

Gontaszewska A. (2001). Evaluation of hydrochemical conditions on the Zawada well field supplying water to the city of Zielona Góra. MA thesis. Arch. Inst. Geologii UAM w Poznaniu. (In Polish).

Górski J. (1981). Formation of groundwater quality in Quaternary deposits in natural conditions and during pumping. IKŚ Warszawa. (In Polish).

Data was made available by the limited liability company "Zielonogórskie Wodociagi i Kanalizacja" Sp. z o.o.

A1.4 SCHRÖGENDORF (A) – SUBTERRANEAN REMOVAL OF IRON AND MANGANESE

Country	Austria
Area of case study	Schrögendorf, Upper Austria
Investigation period	2004-2007
Publication reference	Höfinger, G. (2006), Jung *et al.* (2008)

Drinking water supplier

The Water Association "Fernwasserversorgung Mühlviertel" is a regional water supplier in a mostly rural area in northern Upper Austria.
Population served: about 77,000 people (33 communities).
Annual production: 2,500,000 m^3.

Source and quality of raw water

One of the water sources of Water Association "Fernwasserversorgung Mühlviertel" is the field Schörgendorf with its 4 wells near the village Lugnitz. The water comes from a sandy confined aquifer (Linzer Sand Formation, Molasse zone) with a transmissivity of $1.3 \cdot 10^{-3}$ m^{-2}/s. Pumping tests proved that a maximum of 25 litre/s can be extracted safely from the wells.

The analysis results for Tritium show that the water from the Schörgendorf well field is nearly Tritium-free (results near the detection limit [~0.2 TU]) which gives a retention time of at least 60 years without any noteworthy portion of recent water. This is emphasized by additional carbon dating which revealed a corrected model age of about 5.000 years.

Generally the ground waters of the four wells show a fairly constant chemical composition (see Table A1.4.1).

Table A1.4.1 Initial chemical composition of the groundwater of Schörgendorf well field.

Parameter	Value/range
pH	7.25
Sodium [mg/l]	76–83
Potassium [mg · l]	2.5–2.6
Calcium [mg/l]	32.0–43.0
Magnesium [mg/l]	9.7–11.0
Iron [mg/l]	0.016
Manganese [mg/l]	0.14–0.17
Ammonium [ppm]	1.73
Hydrocarbonates [mg/l]	373–397
Sulfate [mg/l]	7.4–8.2
Nitrate [mg/l]	0.3–1.8
Nitrite [mg/l]	<0.05
Chloride [mg/l]	1.6–2.4
Total Hardness [mg $CaCO_3$/l]	125–150

Despite the reducing conditions only manganese concentrations are above the allowable indicator values, but iron is not. The total hardness below 150 mg $CaCO_3$/l is also remarkable low in the well field. This is caused by ion exchange processes where Calcium is replaced by Sodium. Slightly elevated fluoride and ammonium levels, low nitrate concentrations along with nitrate reduction, and high hydrogen carbonate concentrations are typical for hydro-chemical compositions of deep groundwater in this geological setting.

Problem

The geogenic oxygen reduction of deep groundwater along with high concentrations of manganese and ammonium required a treatment of the extracted groundwater.

Solution

Since reducing environment within deeper aquifers already has been expected, the four wells were planned and drilled with the idea of using them also for in-situ treatment. The management plan foresaw the alternate use of the four wells for treatment and production respectively. The treatment was performed in 4 phases:

(1) Water extraction and aeration phase.
(2) Reinjection phase (reinjection of aerated water).
(3) Reaction phase.
(4) Production phase (partial streams for drinking water production and for aeration).

The behaviour of the relevant parameters in the aquifer were surveyed and modelled to optimize the ratio between reaction phase and production phase (Höfinger 2006). The treatment management plan was based on the modelling and is continually adapted according to the current monitoring results.

After several months of run-in phase the manganese content was reduced to below 0.02 mg/l in all four wells during production phase. The ammonium content was also lowered to 0.2–0.4 mg/l.

Conclusions

- Computer modelling of groundwater system is a useful tool for optimising reaction and production phases during *in situ* iron removal.
- The treatment management plan based on the modelling must be continually adapted to the current monitoring results.

REFERENCES

Höfinger G. (2006). Die in situ Behandlung belasteter Grundwässer. Unpublished master thesis, IV + 77pp., Technical University of Vienna.

Jung M., Spendlingwimmer R. and Höfinger G. (2008). Subterranean Removal of Iron and Manganese in Deep Ground Water – a Case Study in Upper Austria. Metals and Related Substances in Drinking Water, Cost Action 637 II International Conference. Lisbon, Portugal October 2008. Programme and Abstract Book, 39–40.

References

75/440/ECC, Council Directive 75/440/EEC of 16 June 1975 concerning the quality required of surface water intended for the abstraction of drinking water in the Member States. O.J.L 25.7.75, 26–31.

98/83/EC, Council Directive 98/83/EC of 3 November 1998 on the quality of water intended for human consumption. O.J.L 336, 5.12.1998, 32–54.

2000/60/EC, Directive 2000/60/EC of the European Parliament and of the Council of 23 October 2000 establishing a framework for Community action in the field of water policy. O.J.L. 327, 22.12.2000, 1–73.

Abesser C., Robinson R. and Soulsby C. (2006). Iron and manganese cycling in the storm runoff of a Scottish upland catchment. *J Hydrol* **326**, 59–78.

ACC (2004). The use of Phosphates for Notable Water Treatment. American Chemistry Council, www.phosphatesfacts.org.com (accessed 10 February 2013)

AHS (2012). American Hemochromatosis Society. http://www.americanhs.org/faq.htm. (accessed 15 May 2011)

Alvarez-Vazquez L. J., Martinez A., Vazquez-Mendez M. E. and Vilar M. A. (2006). Optimal location of sampling points for river pollution control. *Mathematics and Computers in Simulation*, **71**, 149–160.

Appenzeller B. M., Batte M., Mathieu L., Block J. C., Lahoussine V., Cavard J. and Gatel D. (2001). Effect of adding phosphate to drinking water on bacterial growth in slightly and highly corroded pipes. *Water Res*, **35**(4), 1100–1105.

Arredondo M. and Nunez M. T. (2005). Iron and copper metabolism. *Molecular Aspects of Medicine*, **26**, 313–327.

ASTM D1068-05 Standard Test Methods for Iron in Water. Test Method D-Photometric Bathophenanthroline. doi: 10.1520/D1068–05 (accessed 7 February 2013)

ASTM G31-72 (2004). Standard Practice for Laboratory Immersion Corrosion Testing of Metals. American Society for Testing and Materials. doi: 10.1520/G0031–72R04 (accessed 12 February 2013)

ATSDR (2000). Toxicological Profile for Manganese, U.S. Department of Health and Human, Services, Public Health Service. Agency for Toxic Substances and Disease Registry, Atlanta, GA.

AWWA (2001). Dawn of the Replacement Era. Reinvesting in drinking water infrastructure. American Water Works Association, Denver, CO, pp. 24.

AWWA (2003). Type of pipe material in place for water distribution. *AWWA MainStream*, **47**(1), 7–13.

AWWA (2011). Buried No Longer: Confronting America's Water Infrastructure Challenge. American Water Works Association, Denver, CO, pp. 24.

AWWARF (1990). Chemistry of Corrosion Inhibitors in Potable Water. Report by the AWWA Research Foundation and the American Water Works Association. Denver, Colorado. Benjamin M. M., Reiber S. H., Ferguson J. F., Vanderwerff E. A., and Miller M. W.

AWWARF (1996). Internal corrosion of water distribution systems. AWWARF-DVGW-TZW cooperative research report, Denver, CO, 586 pp.

Baboian R. (ed.). (2002). NACE Corrosion Engineer's Reference Book. 3rd edn, NACE Press, Houston, Texas.

Baboian R. (2005). Corrosion Tests and Standards: Application and Interpretation. 2nd edn, (MNL 20). ASTM International, West Conshohocken, PA.

Barbeau B., Gauthier V., Julienne K. and Carriere A. (2005). Dead-end flushing of a distribution system: Short and long-term effects on water quality. *Journal of Water Supply: Research and Technology—AQUA*, **54**(6), 371–383.

Barcelona M. J., Gibb J. R. and Miller R. A. (1983). A Guide to the Selection of Materials for Monitoring Well Construction and Ground-Water Sampling, Contract Report 327, Illinois State Water Survey, Champaign, IL.

Barceloux D. G. (1999). Manganese. *Clinical Toxicology*, **37**, 293–307.

Baylis J. R. (1926). Prevention of Corrosion and Red Water. *J. Am. Water Works Assoc*, **15**(6), 598.

Benefield L. D., Judkins J. F. and Weand B. L. (1982). Process Chemistry for Water and Waste Water Treatment. Prentice Hall, Inc., USA.

Błaszyk T. and Górski J. (1981). Ground water quality changes during exploitation. *Ground Water*, **19**(1), 28–33.

Bobrowski M. (2002). Sanitary biology. Auxiliary materials for exercises. Białystok Technical University.

Boehrer B. and Schultze M. (2008). Stratification of lakes. *Rev. Geophys.*, **46**, RG2005, doi:10.1029/2006RG000210. (accessed 12 October 2012)

Borch T., Camper A. K., Biederman J. A., Butterfield P. W., Gerlach R. and Amonette J. E. (2008). Evaluation of characterization techniques for iron pipe corrosion products and iron oxide thin films. *Journal of Environmental Engineering ASCE*, **134**(10), 835–844.

Borenstein S. W. (1994). Microbiologically influenced Corrosion Handbook. Woodhead Publishing Ltd, Abington, England.

Bouchard M., Laforest F., Vandelac L., Bellinger D. and Mergler D. (2007). Hair manganese and hyperactive behaviours: pilot study of school-age children exposed through tap water. *Environ Health Perspect*, **115**(1), 122–127.

Bouchard M. F., Sauve S., Barbeau B., Legrand M., Brodeur M-E., Bouffard T. and Limoges E. (2011). Intellectual Impairment in School-Age Children Exposed to Manganese from drinking Water. *Environ Health Perspect*, **119**, 138–143.

Bowman B. and Russell R. (2006). Present Knowledge in Nutrition. 9th edn, Vol. 1, ILSI Press.

Boxall J. B., Saul A. J. and Skipworth P. J. (2001). A novel approach to modelling sediment movement in distribution mains based on particle characteristics. *Water Software Systems*, **1**, 263–273.

Braester C. and Martinell R. (1987). The Vyredox and Nitredoks method in situ treatment of groundwater, International Symposium: Groundwater microbiology problems and biological treatment, Kuopio, Finland.

Broo A. E., Berghult B. and Hedberg T. (1999). Drinking water distribution—The effect of natural organic matter (NOM) on the corrosion of iron and copper. *Water Sci. Technol.*, **40**(9), 17–24.

Broo A. E., Berghult B. and Hedberg T. (2001). Drinking water distribution—Improvements of the surface complexation model for iron corrosion. *Water Sci. Tech. Water Supply*, **1**(3), 11–18.

Bruchet A., Anselme C., Jammes C. and Mallevialle J. (1995). Development of a taste and odor expert system: present state, strengths, limitations and possible future evolution. *Wate. Sci. Tech.* **31**(11), 243–250.

Burrows R., Crowder G. S. and Zhang J. (2000). Utilisation of network modelling in the operational management of water distribution systems. *Urban Water*, **2**, 83–95.

Cannell M. G. R. and Pitcairn C. E. R. (eds) (1993). Impacts of the Mild Winters and Hot Summers in the United Kingdom in 1988–1990. Department of the Environment, HMSO, London, pp. 120–122.

Carlson K. H., Knocke W. R. and Gertig K. R. (1997) Optimizing treatment through Fe and Mn fractionation. *J. Am. Water Works Assoc.*, **89**(4), 2720–2726.

Carriere A., Gauthier V., Desjardins R. and Barbeau B. (2005). Evaluation of loose deposits in distribution systems through unidirectional flushing. *J. Am. Water Works Assoc.*, **97**(9), 82–92.

Casey T. J. (2009). Iron and Manganese in Water Occurrence, Drinking Water Standards, Treatment Options. Aquavarra Research Limited, Blackrock, www.aquavarra.ie (accessed 10 November 2011).

Cerrato J. M., Reyes L. P., Alvarado C. N. and Dietrich A. M. (2006). Effect of PVC and iron materials on Mn(II) deposition in drinking water distribution systems. *Wat. Res.* **40**, 2720–2726.

Charaklis W. G. (1990). Laboratory biofilm reactors. In: Biofilms. W. G. Charaklis and K. C. Marshall (eds) John Wiley Publishers, New York, pp. 55–89.

Chen H. and Copes R. (2011). Manganese in drinking water and intellectual impairment in school-age children. *Environ Health Perspect.*, **119**, a240–a241.

Chiswell B. and Mokhtar M. B. (1986) The speciation of manganese in freshwaters. *Talanta*, **33**(8), 669–677.

Clement J. A. (ed.) (2002). Development of Red Water Control Strategies. American Water Works Association Research Foundation, Denver, CO.

Colter A. and Mahler R. L. (2006). Iron in Drinking Water. University of Idaho PNW 589.

Cornett R. J. and Rigler F. H. (1987). Vertical transport of oxygen into the hypolimnion of lakes. *Canadian Journal of Fisheries and Aquatic Sciences*, **44**(4), 852–858.

Cottis R. A., Graham M. J., Lindsay R., Lyon S. B., Richardson J. A., Scantlebury J. D. and Scot F. H. (eds) (2010). Shreir's Corrosion. 4th edn. Elsevier, Amsterdam, The Netherlands. ISBN 9780444527882.

Dąbrowski W. (2008). Cement mortar. *Instal* **5**, 39–41. (In Polish)

Dąbrowski W. (2011). Pipe lining with cement mortar compared with other renovation technologies. *BMP Ochrona Srodowiska*, **3**, 32–36. (In Polish)

Dallman P. (1986). Biochemical basis for the manifestations of iron deficiency. *Ann. Rev. Nutr*, **6**, 13–40.

Davies C., Fraser D. L., Hertzler P. C. and Jones R. T. (1997). US EPA's infrastructure needs survey. *J. Am. Water Works Assoc.*, **89**, 12–30.

Davison W. (1993). Iron and manganese in lakes. *Earth-Science Reviews*, **34**(2), 119–163.

Denczew S. and Królikowski A. (2002). Fundamentals of Modern Maintenance of Water Distribution Networks and Sewage Systems, Arkady, Warsaw (In Polish).

Deshommes E., Laurent Laroche L., Nour S., Cartier C. and Prevost M. (2010). Source and occurrence of particulate lead in tap water. *Wat. Res.* **44**, 3734–3744.

DHS (1992). The Use of Polyphosphates as Sequestering Agent for Iron and Manganese Controll. American Department of Health Services Guidelines, Washington, DC.

Dodrill D. and Edwards M. (1995). Corrosion control on the basis of utility experience. *J. Am. Water Works Assoc.*, **87**(7), 74–85.

Doria M. F., Pidgeon N. and Hunter P. R. (2009). Perceptions of drinking water quality and risk and its effect on behaviour: A cross-national study. *Science of the Total Environment*, **407**, 5455–5464.

DVGW nr W216, Versorgung mit unterschiedlichen Trinkwässern, August, 2004 (in German).

Dvorak B. I., Prasai G., Skipton S. O. and Woldt W. (2007). Drinking Water: Iron and Manganese. Water Resources Management. Drinking Water. NIANR. University of Nebraska – Lincoln Extention, Institute of Agriculture and Natural Resources. http://www.ianrpubs.unl.edu/epublic/pages/index.jsp (accessed 10 May 2011)

Dzombak D. A. and Morel F. F. M. (1990). Surface Complexation Modeling: Hydrous Ferric Oxide. Wiley-Interscience, New York.

Edwards M. (2004). Controlling corrosion in drinking water: grand challenge for the 21st century. *Water Sci. Tech.*, **2**(49), 1–8.

Eisnor J. D. and Gagnon G. A. (2003). A framework for the implementation and design of pilot-scale distribution systems. *J. Water Supply Res. Technol. – Aqua*, **57**(7), 501–520.

Emerson D. and Revsbech N. P. (1994a). Investigation of an iron-oxidizing microbial mat community located near Aarhus, Denmark: field studies. *Appl. Environ. Microbiol.*, **60**, 4022–4031.

Emerson D. and Revsbech N. P. (1994b). Investigation of an iron-oxidizing microbial mat community located near Aarhus, Denmark: laboratory studies. *Appl. Environ. Microbiol.*, **60**, 4032–4038.

Engelhardt M. O., Skipworth P. J., Savic D. A., Saul A. J. and Walters G. A. (2000). Rehabilitation strategies for water distribution networks: a literature review with UK perspective. *Urban Water*, **2**, 153–170.

Environment Canada (1984). Detailed surface water quality data, Northwest Territories 1980–1981, Alberta 1980–1981, Saskatchewan 1980–1981, Manitoba 1980–1981. Inland Waters Directorate (1984).

Epke E. M. and Lawless H. T. (2007). Retronasal smell and detection thresholds of iron and copper salts. *Physiology & Behavior*, **92**, 487–491.

Evins C., Liebeschuetz J. and Williams S. M. (1990). Aesthetic Water Quality Problems in Distribution systems. A *S*ource *D*ocument for the *W*ater *M*ains *R*ehabilitation *M*anual. Water Research Centre, Swindon, UK.

Fahey T. J., Hill M. O., Stevens P. A., Hornung M. and Rowland P. (1991a). Nutrient accumulation in vegetation following conventional and whole-tree harvest of Sitka spruce plantations in north Wales. *Forestry*, **64**, 271–288.

Fahey T. J., Stevens P. A., Hornung M. and Rowland P. (1991b). Decomposition and nutrient release from logging residue following conventional harvest of Sitka spruce in north Wales. *Forestry*, **64**, 289–301.

Faust S. D. and Aly O. M. (1997). *Chemistry of Water Treatment*. Ann Arbor Press, Michigan.

Finley J. W. (1999). Manganese absorption and retention by young women is associated with serum ferritin concentration. *Am J Clin Nutr*, **70**, 37–43.

Florence T. M. (1982). The speciation of trace elements in water. *Talanta*, **29**, 345–364.

Fraga C. G. (2005). Relevance, essentiality and toxicity of trace elements in human health. *Molecular Aspects of Medicine*, **26**, 235–244.

Fraga C. G. and Oteiza P. I. (2002). Iron toxicity and antioxidant nutrients. *Toxicology*, **180**, 23–32.

Frankel G. S. (2008). Electrochemical Techniques in corrosion: Status, limitations, and needs. *Journal of ASTM International*, **5**(2), 1–27.

Freeland-Graves J. H., Bales C. W. and Behmardi F. (1987). Manganese requirements of humans. In: Nutritional Bioavailability of Manganese. C. Kies , (ed.), American Chemical Society, Washington, D.C, pp. 90–104.

Friedman M., Kirmeyer G. J. and Antoun E. (2002). Developing and implementing a distribution system flushing program. *J. Am. Water Works Assoc.*, **94**(7), 48–56.

Gancel G., Moratzavi I. and Piller O. (2006). Coupled numerical simulations and sensitivity assessment for quality modelling for water distribution systems. *Applied Mathematics Letters*, **19**, 1313–1319.

Gault A. G., Ibrahim A., Langley S., Renaud R., Takahashi Y., Boothman C., Lloyd J. R., Clark I. D., Ferris F. G. and Fortin D. (2011). Microbial and geochemical features suggest iron redox cycling within bacteriogenic iron oxide-rich sediments. *Chemical Geology*, **281**, 41–51.

Gauthier V., Barbeau B., Millette R., Block J.-C. and Prevost M. (2001). Suspended particles in the drinking water of two distribution systems. *Water Sci. Tech.: Water Supply*, **1**(4), 237–245.

Gebbie P. (2000) Water stability – what does it mean and how Do you measure it? 63rd Annual Water Industry Engineers and Operators' Conference Civic Centre, Warrnambool 6 and 7 September, 2000.

Giani R. (2008). Possible Influence of Galvanized Plumbing in Elevated Lead Levels. Conference Proceedings, AWWA Annual Conference and Exposition, 2008.

Gianutsos G., Morrow G. R. and Morris J. B. (1997). Accumulation of manganese in rat brain following intranasal administration. *Applied Toxicol*, **37**, 102–105.

Gontaszewska A. (2001). Evaluation of Hydrochemical Conditions on the Zawada well Field Supplying Water to the City of Zielona Gora. MA thesis. Arch. Inst. Geologii UAM w Poznaniu. (In Polish).

Górski J. (1981). Formation of Groundwater Quality in Quaternary Deposits in Natural Conditions and During Pumping. Wyd. IKŚ, Warszawa, (In Polish).

Górski J. (1989). The main hydrochemical problems of Cainozoic aquifers located in Central Wielkopolska (Great Poland). Zesz. Nauk AGH Kraków, (In Polish).

Górski J. (2004). Evaluation of groundwater quality and anthropogenic contamination. In: Determination of Groundwater Resources for Wells and Well-Filds, S. Dąbrowski, J. Górski, J. Kapuściński, J. Przybyłek and A. Szczepański (eds), Ministry of Environment, Warszawa, pp. 60–77, (In Polish).

Górski J. (2011). Treatment of groundwater in aquifer. In: Treatment of Water. Physical, Chemical and Biological Processes, J. Nawrocki (ed.), Wyd. Nauk. PWN, Warszawa, pp. 316–356, (In Polish).

Greger J. L. (1999). Nutrition versus toxicology of manganese in humans: Evaluation of potential biomarkers. *NeuroToxicol*, **20**, 205–212.

Griffin A. E. (1960). Significance and removal of manganese in water supplies. *J. Am. Water Works Assoc.* **52**, 1326–1334.

Grochulska-Segal E. M. and Sozański M. M. (1984). Selected issues of iron and manganese removal technology from groundwater. Environmental Pollution Control. Journal of Polish Sanitary Engineers' Association, **134**, XVIII, Wroclaw, Poland.

Grossblatt J. (1998). Phosphate blend controls corrosion at lower dosage rates. *Water World*, **14**(10), 18.

Guilarte T. R. (2010). Manganese and Parkinson's disease: a critical review and new findings. *Env. Health Perspect*, **118**(8), 1071.

Guthrie J., Battat B. and Grethlein C. (2002). Accelerated corrosion testing. *The AMPTIAC Quarterly*, **6**(3), 11–15.

Hadzilacos T., Kalles T., Preston D., Melbourne N., Camarinopoulos P., Eimermacher L., Kallidromitis M., Frondistou-Yannas V. and Saegrov S. (2000). UtilNets: a water mains rehabilitation decision-support system. *Computers, Environment and Urban Systems*, **24**, 215–232.

Hallberg R. O. and Martinell R. (1976). Vyredox – in situ purification of groundwater. *Ground Water*, **14**(2), 88–93.

Hamilton-Taylor J., Davison W. and Morfett K. (1996). The biogeochemical cycling of Zn, Cu, Mn, and dissolved organic C in a seasonally anoxic lake. *Limnol. Oceanogr*, **41**, 408–418.

Hardie A. M., Heal K. V. and Lilly A. (2007) The influence of pedology and changes in soil moisture status on manganese release from upland catchments: soil core laboratory experiments. *Water Air Soil Pollut.*, **182**, 369–382, doi: 10.1007/s11270-007-9348-6. (accessed 17 October 2012)

Hater W., Wagner I. and Schmidt P. (2002). Experiences with application of inhibitors to reduce heavy metal uptake of drinking water, www.environmental-expert.com./technology/henkel (accessed 14 May 2012)

HDR (2009). An Analysis of the Correlation between Lead Released from Galvanized Iron Piping and the Contents of Lead in Drinking Water. Summary report. Bellevue, WA.

Heal K. (2001). Manganese and land-use in upland catchments in Scotland. *The Science of the Total Environment*, **265**, 169–179.

Health Canada (1979). Odour. www.hc-sc.gc.ca (accessed 12 November 2012)

Hem J. D. (1985). Study and interpretation of the chemical characteristics of natural water. US Geological Survey Water-Supply Paper 2254. Washington, USA.

Hoekstra E. J., Aertgeerts R., Bonadonna L., Cortvriend J., Drury D., Goossens R., Jiggins P., Lucentini L., Mendel B., Rasmussen S., Tsvetanova Z., Versteegh A. and Weil M. (2008). The advice of the Ad-Hoc Working Group on Sampling and Monitoring to the Standing Committee on Drinking Water concerning sampling and monitoring for the revision of the Council Directive 98/83/EC. Office for Official Publications of the European Communities, Luxembourg, EUR 23374 EN.

Höfinger G. (2006). Die in situ Behandlung belasteter Grundwässer. – Unpublished master thesis, IV+77pp., Technical University of Vienna.

Holden B., Greetham M., Croll B. T. and Scutt J. (1995). The effect of changing inter process and final disinfection reagents on corrosion and biofilm growth in distribution pipes. *Wat. Sci.Tech.*, **32**(8), 213–220.

Hongve D. (1997). Cycling of iron, manganese, and phosphate in a meromictic lake. *Limnol. Oceanogr.*, **42**, 635–647.

Huang Y. H. and Zhang T. C. (2005). Effects of dissolved oxygen on formation of corrosion products and concomitant oxygen and nitrate reduction in zero-valent iron systems with or without aqueous Fe^{2+}. *Wat. Res.*, **39**, 1751–1760.

Hudak P. F. (2005). Principles of Hydrogeology. 3rd edn, CRC Press.

Husband P. S. and Boxall J. B. (2011). Asset Deterioration and discolouration in water distribution system. *Wat. Res.*, **45**, 113–124.

Husband P. S., Boxall J. B. and Saul A. J. (2008). Laboratory studies investigating the processes leading to discolouration in water distribution networks. *Wat. Res.*, **42**, 4309–4318.

Huser B. J., Kohler S. J., Wilander A., Johansson K. and Folster J. (2011). Temporal and spatial trends for trace metals in streams and rivers across Sweden (1996–2009). *Biogeosciences*, **8**, 1813–1823.

Imran S. A., Dietz J. D., Mutoti G., Taylor J. S., Randall A. A. and Cooper C. D. (2005a). Red Water release in Drinking Water Distribution System. *J. Am. Water Works Assoc.*, **97**(9), 93–100.

Imran S. A., Dietz J. D., Mutoti M. G., Taylor J. S. and Randall A. A. (2005b). Modified Larson's Ratio Incorporating Temperature, Water Age, and Electroneutrality Effects on Red Water Release. *Journal of Environmental Engineering*, **131**(11), 1514–1520.

IOM (2001). Dietary Reference Intakes for Vitamin A, Vitamin K, Arsenic, Boron, Chromium, Copper, Iodine, Iron, Manganese, Molybdenum, Nickel, Silicon, Vanadium, and Zinc. Institute of Medicine, Food and Nutrition Board, Washington, DC, National Academy Press.

IOM (2002). Dietary Reference Intakes: Vitamin A, Vitamin K, Arsenic, Boron, Chromium, Copper, Iodine, Iron, Manganese, Molybdenum, Nickel, Silicon, Vanadium, and Zinc. Institute of Medicine. Food and Nutrition Board, Washington DC, National Academy Press.

Ishigaki T., Nakanishi A., Tateda M., Ike M. and Fujita M. (2005). Bioleaching of metal from municipal waste incineration fly ash using a mixed culture of sulfur-oxidizing and iron-oxidizing bacteria. *Chemosphere* **60**, 1087–1094.

ISO 5667-1:2006. Water quality – Sampling – Part 1: Guidance on the design of sampling programmes and sampling techniques.

ISO 5667-11:2009. Water quality – Sampling – Part 11: Guidance on sampling of groundwaters.

ISO 5667-14:2004. Water quality – Sampling – Part 14: Guidance on quality assurance of environmental water sampling and handling.

ISO 5667-20:2008. Water quality – Sampling – Part 20: Guidance on the use of sampling data for decision making – Compliance with thresholds and classification systems.

ISO 5667-22. 2010. Water quality – Sampling – Part 22: Guidance on the design and installation of groundwater monitoring points.

ISO 5667-23.2011. Water quality – Sampling – Part 23: Guidance on passive sampling in surface waters.

ISO 5667-3:2003. Water quality – Sampling – Part 3: Guidance on the preservation and handling of water samples.

ISO 5667-4:1987. Water quality – Sampling – Part 4: Guidance on sampling from lakes, natural and man-made.

ISO 5667-5:2006. Water quality – Sampling – Part 5: Guidance on sampling of drinking water from treatment works and piped distribution systems.

ISO 5667-6:2005. Water quality – Sampling – Part 6: Guidance on sampling of rivers and streams.

ISO 6332:1988 Water quality – Determination of iron – Spectrometric method using 1,10-phenanthroline. http://www.iso.org/ (accessed 15 November 2012)

ISO 6333:1986 Water quality – Determination of manganese – Formaldoxime spectrometric method. http://www.iso.org/ (accessed 15 November 2012)

IWA (2012a). Code of Practice for the Internal Corrosion Control of Water Supply Systems. IWA Publishing, London, UK. ISBN 9781780404554.

IWA (2012b). Best Practice Guide on the Removal of Metals from Drinking Water by Treatment. IWA Publishing, London, UK. ISBN: 9781843393849

IWA (2012c). Best Practice Guide for the Sampling of Metals in Drinking Water. IWA Publishing, London, UK. ISBN 978 1843393832.

Jeż-Walkowiak J. (2000). Intensification of Iron and Manganese Removal from Groundwater in Oxidative Beds of Rapid Filters. Ph.D. thesis, Poznan University of Technology, Institute of Environmental Engineering.

Jung M., Spendlingwimmer R. and Höfinger G. (2008). Subterranean Removal of Iron and Manganese in Deep Ground Water – a Case Study in Upper Austria.- Metals and Related Substances in Drinking Water, Cost Action 637 II International Conference. Lisbon, Portugal, October 2008. Programme and Abstract Book, 39–40.

Juyun L. J. and Lawless H. T. (2005). Oral sensations from iron and copper sulfate. *Physiology & Behavior* **85**, 308–313.

Kalavska D. (1991). Determination of Mn(II) in tap water. *Int. Journal Env. Anal. Chem.*, **45**, 159–167.

Kleiner Y. and Rajani B. (2001). Comprehensive review of structural deterioration of water mains: statistical models. *Urban Water*, **3**, 157–176.

Kolonko A. (1998). Classification and characteristics of underground pipe-work systems cleaning methods. Gaz, Woda i Technika Sanitarna, **2**, 74–77. (In Polish).

Krasner S. W., Hwang C. J. and McGuire M. J. (1983). A Standard Method for Quantification of Earthy-Musty Odorants in Water, Sediments and Algal Cultures. *Water Sci. Tech.*, **15**, 127–138.

Kritzberg E. S. and Ekström S. M. (2012). Increasing iron concentrations in surface waters – a factor behind brownification? *Biogeosciences*, **9**, 1465–1478.

Kuiters A. T. and Mulder W. (1993). Water-soluble organic matter in forest soils 1. Complexing properties and implications for soil equilibria. *Plant Soil*, **152**, 215–224.

Kulbik M. (2002). The influence of hydraulic conditions within distribution network on the secondary contamination of drinking water. V International Conference: Water Supply and Water Quality, Gdansk-Poznan, 2002. (In Polish).

Langelier W. F. (1936). The analytical control of anti-corrosion water treatment. *J. Am. Water Works Assoc.*, **8**, 500–1521.

Laohaudomchok W., Lin X., Herrick R. F., Fang S. C., Cavallari J. M., Christiani D. C. and Weisskopf M. G. (2011). Toenail, blood and urine as biomarkers of manganese exposure. *J Occup Environ Med*, **53**(5), 506–10.

Larson T. E. (1960). Loss in pipeline carrying capacity due to corrosion and tuberculation. *J. Am. Water Works Assoc.*, **52**(10), 1263–1270.

Larson T. E. (1939). Bacteria, corrosion and red water. *J. Am. Water Works Assoc.*, **31**(7), 1186–1196.

Larson T. E. and King R. M. (1954). Corrosion by Water at Low Flow Velocity. State Of Illinois, Department of Registration and Education, Urbana, Illinois.

Larson T. E. and Skold R. V. (1957). Corrosion and tuberculation of cast iron. *J. Am. Water Works Assoc.*, **49**, 1294–1301.

Larson T. E. and Skold R. V. (1958). Laboratory studies relating mineral quality of water to corrosion of steel and cast iron. *Corrosion*, **14**(8), 285–288.

Larson T. E. (1966). Chemical corrosion control. *J. Am. Water Works Assoc.*, **49**(12), 1581.

Laxen D. P. H., Davison W. and Woof C. (1984). Manganese chemistry in rivers and streams. *Geochim Cosmochim Acta*, **48**, 2107–2111.

LaZerte B. D. and Burling K. (1990). Manganese speciation in dilute waters of the Precambrian Shield, Canada. *Wat. Res.*, **24**, 1097–1101.

Le Gat Y. and Eisenbeis P. (2000). Using maintenance records to forecast failures in water network. *Urban Water*, **2**, 173–181.

LeChevallier M. W., Babcock T. M. and Lee R. G. (1987). Examination and characterization of distribution system biofilms. *Appl. Environ. Microbiol.*, **53**(12), 2714–2724.

Lee D. Y., Kim W. C. and Kim J. G. (2012). Effect of nitrite concentration on the corrosion behaviour of carbon steel pipelines in synthetic tap water. *Corrosion Science*, **64**, 105–114.

Letterman R. D. (ed.) (1999). Water Quality and Treatment. A Handbook of Community Water Supplies. 5th edn, American Water Works Association. McGraw-Hill, Inc., New York, US.

Levallois P., Grondin J. and Gingras S. (1999). Evaluation of consumer attitudes on taste and tap water alternatives in Quebec. *Journal of Environmental Management*, **70**, 363–373.

Lin S. D. (1977). Tastes and Odors in Water Supplies— A Review. Illinois State Water Survey Urbana.

Lipponen M. T. T., Martikainen P. J., Vasara R. E., Servomaa K., Zacheus O. and Kontro M. H. (2004). Occurrence of nitrifiers and diversity of ammonia-oxidizing bacteria in developing drinking water biofilms. *Water Res.*, **38**, 4424–4434.

Lipponen M. T. T., Suutari M. H. and Martikainen P. J. (2002). Occurrence of nitrifying bacteria and nitrification in finnish drinking water distribution system. *Water Res.*, **36**, 4319–4329.

Little D. I. and McFadzean C. J. (1991). Bradan water treatment works extension and upgrading. In: Water and Environmental Management, M. D. F. Haigh and C. P. James (eds), Ellis Horwood, Chichester, UK, pp. 178–188.

Ljung K. and Vahter M. (2007). Time to Re-evaluate the guideline value for manganese in drinking water. *Environ Health Perspect.*, **115**(11), 1533–1538.

Loewenthal R. E. and Marais G. V. R. (1982). Carbonate Chemistry of Aquatic Systems. Ann Arbor Science. *AIChE J.*, **25**(557), pp. 433 +xi, $22.50. doi: 10.1002/aic.690250328. (accessed 25 august 2011)

Łomotowski J. (2007). Reasons for Water Quality Changes in Water Distribution Systems. Polish Academy of Sciences, Systems Research Institute, Warsaw. (In Polish).

Lonnerdal B. (1994). Manganese nutrition of infants. Chapter 11. In: Manganese in Health and Disease, Klimis-Tavantzis (ed.), CEC Press, Boca Raton, Florida, 212, pp. 175–192.

Lueschow L. A. and Mackenthun K. M. (1962). Detection and Enumeration of Iron Bacteria in Municipal Water Supplies. *J. Am. Water Works Assoc.*, **54**(6), 751–756.

Lytle D. A., Magnuson M. L. and Snoeyink V. L. (2004). Effect of oxidants on the properties of Fe(III) particles and suspension formed from the oxidation of Fe(II). *J. Am. Water Works Assoc.*, **96**(8), 112.

Lytle D. A., Sarin P. and Snoeyink V. L. (2005). The effect of chloride and orthophosphate on the release of iron from a cast iron pipe section. *Journal of Water Supply: Research and Technology—AQUA*, **54**(5), 267–281.

Macioszczyk A. (1987). Hydrogeochemistry. Wyd. Geol, Warszawa. (In Polish).

Maj T. (1999). Directional flushing as one of the most important parts of maintenance programmes in drinking water distribution system. *Gaz, Woda i Technika Sanitarna*, **10**, 360–361 (In Polish).

Manganese Health Research Program (MHRP) by the International Manganese Institute (IMnI). http://www.manganese-health.org/about_us/healtheffects (accessed 25 May 2011)

Mansfeld F. (1976). The polarization resistance technique for measuring corrosion currents. In: Advances in Corrosion Science and Technology, M. G. Fontana and R. W. Staehle (eds), Plenum, New York.

Matthews J. C., Selvakumar A., Sterling R. L. and Condit W. (2012). Innovative Rehabilitation Technology Demonstration and Evaluation Program, *Tunneling and Underground Space Technology*, http://dx.doi.org/10.1016/j.tust.2012.02.003.

McFadden M., Gianni R., Kwan P. and Reiber S. H. (2011). Contributions to drinking water lead from galvanized iron corrosion scales. *J. Am. Water Works Assoc.*, **103**(4), 76–89.

McGivney W. T. and Kawamura S. (2008). Cost Estimating Manual for Water Treatment Facilities. John Wiley & Sons, Inc., Hoboken, New Jersey, 244 pp.

McNeill L. S. and Edwards M. (2001). Iron pipe corrosion in distribution systems. *J. Am. Water Works Assoc.*, **93**(7), 88–100.

McNeill L. S. and Edwards M. (2002). The importance of temperature in assessing iron pipe Corrosion in water distribution systems. *Environmental Monitoring and Assessment*, **77**, 229–242.

McNeill L. S. and Edwards M. (2000). Phosphate inhibitors and red water in stagnant iron pipes. *Journal of Enviromental Engineering*, **126**(12), 1096–1102.

Moglia M., Perez P. and Burn S. (2010). Modelling an urban water system on the edge of chaos. *Environmental Modelling and Software*, **25**, 1528–1538.

Morgan J. J. (1967a). Application and Limitation of Chemical Thermodynamics in Water Systems w Equilibrium Concepts in Natural Water Systems. American Water Society, Washington.

Morgan J. J. (1967b). Chemical Equilibria and Kinetic Properties of Manganese in Natural Waters w Principles and Applications in Water Chemistry. John Wiley & Sons, Inc., New York.

Mouchet P. (1992). From conventional to biological removal of iron and manganese in France. *J. Am. Water Works Assoc.*, **84**(4), 158–167.

Murdoch F. and Smith P. G. (1999). Formation of manganese micro-nodules on water pipeline materials. *Wat. Res.*, **33** (12), 2893–2895.

MWH (2005). Water Treatment Principles and Design. 2nd edn, Revised by J.C. Crittenden R.R. Trussell D.W. Hand K. J. Howe and G. Tchobanoglous, John Wiley & Sons, Inc., Hoboken, NJ.

NACE International Standard Test Method M0169-2000, Laboratory Corrosion Testing of Metals. ISBN: 1-57590-098-X.

NACE TMO 169-76 Laboratory Corrosion Testing of Metals for the Process Industries. NACE International, Houston, Texas.

Naddeo V., Zarra T. and Belgiorno V. (2007). Optimization of sampling frequency for river water quality assessment according to Italian implementation of the EU Water Framework Directive. *Environmental Science & Policy*, **10**, 243–249.

Nawrocki J., Raczyk-Stanisawska U., Świetlek J., Olejnik A. and Sroka M. (2010). Corrosion in a distribution systems: steady water and its composition. *Wat. Res.*, **44**(6), 1863–1872.

Neal C., House W. A., Leeks G. J. L., Whitton B. A. and Williams R. J. (2005). The water quality of UK rivers entering the North Sea. *The Science of the Total Environment*, **251/252**, 5–8.

Neal C., Lofts S., Evans C. D., Reynolds B., Tipping E. and Neal M. (2008). Increasing iron concentrations in UK upland waters. *Aquat. Geochem.*, **14**, 263–288.

NIEHS (2010). National Institute of Environmental Health Sciences. National Institute of Health US department of Health and Human Sciences. Manganese & Brain Damage. NIEHS press, Research Triangle Park, NC.

Nielsen D. M. (ed.) (2006). Practical Handbook of Environmental Site Characterization and Ground-water Monitoring. Second edition. Taylor & Francis, New York.

Odell L. H., Kirmeyer G. J., Wilczak A., Jacangelo J. G., Marcinko J. P. and Wolfe R. L. (1996). Controlling nitrification in chloraminated systems. *Jour. AWWA*, **88**(7), 86–98.

Olańczuk-Neyman K. (1988). Bacteria oxidizing iron and their occurrence in groundwater, Scientific Notebooks of Gdansk Technical University 415, Hydro-engineering, No. XXVIII, Gdansk, Poland.

Olańczuk-Neyman K. (2001). Microorganismsm in Shaping the Quality and Groundwater Treatment. Publishing House of Gdańsk Technical University, Gdańsk, Poland.

Parker L. (1994). The effects of ground-water sampling devices on water quality: a literature review. *Ground-Water Monitoring and Remediation*, **14**(2), 130–141.

Pattison V. A., Butcher D. P., Labadz J. C. and Shacklock J. (1995). Water Discolouration and the Role of Reservoir. *Phys. Chem. Earth*, **20**(2), 175–181.

Perchuć M. (1996). Problems with groundwater treatment, Conference Mat. of I Scientific-Technical Conference on: Treatment of Groundwater – studies, design and exploitation, Warsaw, Poland.

Peryam D. R. (1958). Sensory difference tests. *Food Technol.*, **12**, 231–236.

Piechurski F. and Dzikowski T. (2012). Cost estimation of water mains rehabilitation and replacement, (In Polish), Nowoczesne Budownictwo Inżynieryjne, pp. 31–34.

Pintar K. D. M. and Slawson R. M. (2003). Effect of temperature and disinfection strategies on ammonia-oxidizing bacteria in a bench-scale drinking water distribution system. *Wat. Res.*, **37**(8), 1805–1817.

Pohlmann K. F., Icopini G. A., McArthur R. D. and Rosal C. G. (1994). Evaluation of Sampling and Field-Filtration Methods for Analysis of Trace Metals in Ground Water. U.S. Environmental Protection Agency, Office of Research and Development, Las Vegas, NV.

Postawa A., Jeż-Walkowiak J., Pruss A. and Wątor K. (2011). Arsenic in groundwaters near Lublin. Bulletin of Polish Geology Institute. *Hydrogeology*, **445**(12/2), 495–503.

Prakash T. M., Scholze R. J., Neff C. H., Maloney S. W., Heath M. and Smith E. D. (1988). *Development of the Pipe Loop System for F etermining Effectiveness of Corrosion Control Chemicals in Potable Water Systems.* USA-CERL technical report N-88/12.

Pruss P. and Pruss A. (2010). Water Treatment Plant of the Underground of Technological Research to Start. GWITS, Nov 2010, Poznan, Poland (in Polish).

Pullin M. J. and Cabaniss S. E. (2003). The effects of pH, ionic strength, and iron–fulvic acid interactions on the kinetics of nonphotochemical iron transformations. I. Iron(II) oxidation and iron(III) colloid formation. *Geochimica et Cosmochimica Acta*, **67**(21), 4067–4077.

Reiber S., Ferguson J. F. and Benjamin M. M. (1988). An improved method for corrosion-rate measurement by weight loss. *J. Am. Water Works Assoc.* **80**(11), 41–46.

Reiber S. H., Ryder R. A. and Wagner I. (1996). Corrosion assessment technologies. In: Internal Corrosion of Water Distribution Systems, 2nd edn, American Water Works Association Research Foundation and DVGW Technologiezentrum Wasser, Denver, CO, pp. 445–486.

Revie W. H. (ed.) (2000). Uhlig's Corrosion Handbook. 2nd edn, John Wiley & Sons, Inc., Hoboken, New Jersey.

Revie W. H. and Uhlig H. H. (2008). Corrosion and Corrosion Control. An Introduction to Corrosion Science and Engineering. 4th edn, John Wiley & Sons, Inc., Hoboken, New Jersey.

Roberge P. R. (2007) Corrosion Inspection and Monitoring. John Wiley & Sons Inc., Hoboken, New Jersey.

Rogers H. R. (2002) Review of test methods and criteria for assessing taste and off-taste. Final Report to DEFRA Report No: DEFRA 5013/1.

Rosborg I. (2008). High iron concentrations in drinking water – a potential health risk? Metals and Related Substances in Drinking Water, Cost Action 637 II International Conference. Lisbon, Portugal, October 2008, pp. 286–297.

Ryan A. (1997). Iron-deficiency anemia in infant development: implications for growth, cognitive development, resistance to infection, and iron supplementation. *Yearbook of Physical Anthropology*, **40**, 25–62.

Rybicki S. M. (2010). Change of construction cost in Polish municipal infrastructure (In Polish) Report to Ministry Of Environment, Republic of Poland.

Ryznar J. W. (1944). A New Index for Determining Amount of Calcium Carbonate Scale Formed by a Water. *J. Am. Water Works Assoc.*, **36**, 472.

Saegrov S., Melo Baptista J. F., Conroy P., Herz R. K., LeGauffre P., Moss G., Oddevald J. E., Rajani B. and Schiatti M. (1999). Rehabilitation of water networks. Survey of research needs and on-going efforts. *Urban Water*, **1**, 15–22.

Sander A., Berghult B., Ahlberg E., Elfstrom Broo A., Lind Johansson E. and Hedberg T. (1997). Iron corrosion in drinking water distribution systems—surface complexation aspects. *Corrosion Science*, **39**(1), 77–93.

Sarin (2002) iron release from corrosion scales in old iron/steel Drinking water distribution pipes. PhD thesis. University of Illinois at Urbana–Champaign.

Sarin P., Snoeyink V. L., Bebee J., Kriven W. M. and Clement J. A. (2001). Physico-chemical characteristics of corrosion scales in old iron pipes. *Wat. Res.*, **35**(12), 2961–2969.

Sarin P., Snoeyink V. L., Bebee J., Jim K. K., Beckett M. A., Kriven W. M. and Clement J. A. (2004b). Iron release from corroded iron pipes in drinking water distribution systems: effect of dissolved oxygen. *Wat. Res.*, **38**, 1259–1269.

Sarin P., Snoeyink V. L., Clement J. A. and Kriven W. M. (1995). Iron release from corroded unlined cast iron pipes: effect of pH, alkalinity and orthophosphates in a soft New England water. *J. Am. Water Works Assoc.* **11**, 85–96.

Sarin P., Snoeyink V. L., Lytle D. A. and Kriven W. M. (2004a). Iron corrosion scales: model for scale growth, iron release and colored water formation. *Journal of Environmental Engineering*, **130**, 365–373.

Schofield T., Perkins R. and Simms J. S. (1991). Frankley water-treatment works development: pilot-scale studies. *JIWEM*, **5**, 370–380.

Scully J. R. (2000). Polarization resistance method for determination of instantaneous corrosion rates. *Corrosion*, **56**, 199–218.

Seeman H. B. (1996). Temporal Variability of Arsenic Transport into the Upper Mystic Lake. MS thesis, Massachusetts Institute of Technology.

Seth A., Bachmann R., Boxall J., Saul A. J. and Edyvean R. (2003). Characterisation of materials causing discolouration in potable water systems. In: 1st IWA Conference on Scaling and Corrosion in Water and Wastewater Systems, Cranfield.

Setiadi Y., Tanyimboh T. T. and Templeman A. B. (2005). Modelling errors, entropy and the hydraulic reliability of water distribution systems. *Advances in Engineering Software*, **36**, 780–788.

Shalini Chaturvedi Pragnesh N. Dave. (2012). Removal of iron for safe drinking water. *Desalination* **303**, 1–11.

Shreir L. L., Jarman R. A. and Burstein G. T. (1994). Corrosion. 3rd edn, Volumes **1–2**, Elsevier, Oxford, England.

Siepak M., Novotný K., Vaculovič T., Górski J. and Przybyłek J. (2010). Variability of chemical composition of groundwater at the Miocene Aquifer in the Poznań-Gostyń fault graben region (Poland). *Polish Geological Institute Bulletin*, **441**, 145–156.

Sikora A. and Błaszczyk M. K. (2009). Prokaryotes reducing Fe(III). *Postępy Mikrobiologii*, **48**, 105–124. (In Polish).

Skotak K., Swiatczak J., Witczak S. and Postawa A. (2007). Metals and related substances in raw and tap waters in Poland. In: Metals and Related Substances in Drinking Water, Cost Action 637 International Conference. Antalya, Turkey, 24–26 October, pp. 52–60.

Sly L., Hodgekinson M. and Arunpairojana V. (1990). Deposition of manganese in a drinking water distribution system. *Appl. Environ. Microbiol.*, **56**(3), 628–639.

Smargiassi A. and Mutti A. (1999). Peripheral biomarkers and exposure to manganese. *Neurotoxicology*, **20**(2–3), 401–406.

SMEWW 2150 (2011). Standard Methods for the Examination of Water and Wastewater; method 2150: Odor. http://standardmethods.org/ (accessed 10 December 2012)

SMEWW 2170 (2011). Standard Methods for the Examination of Water and Wastewater; method 2170: Flavor Profile Analysis. http://standardmethods.org/ (accessed 10 December 2012)

SMEWW 3500. Standard Methods for the Examination of Water and Wasteewater. Method 3500b -Fe Iron. Phenanthroline Method. http://standardmethods.org (accessed 10 December 2012)

SMEWW 3500. Standard Methods for the Examination of Water and Wasteewater. Method 3500b -Mn Manganese. Persulfate Method. http://standardmethods.org (accessed 10 December 2012)

Smith S. E., Ta T., Holt D. M., Delanoue A., Colbourne J. S., Chamberlain A. H. L. and Lloyd B. J. (1999). A pipeline testing facility for the examination of pipe-wall deposits and red-water events in drinking water. *J.CIWEM*, **13**, 7–15.

Solo-Gabriele H. M. and Perkins F. E. (1997b). Metal transport within a small urbanized watershed. *J. Irrig. Drain. Eng. ASCE*, **123**, 114–122.

Solo-Gabriele H. M. and Perkins F. E. (1997a). Watershed-specific model for streamflow, sediment and metal transport. *J. Environ. Eng. ASCE* **123**, 61–70.

Sommerfield E. O. (1999). Iron and Manganese Removal Handbook. American Water Works Association, Denver.

Sontheimer H., Kolle W. and Snoeyink V. L. (1981). The siderite model of the formation of corrosion-resistant scales. *J. Am. Water Works Assoc.*, **71**(11), 572–579.

Sozański M. and Huck P. M. (2007). Experimental Investigations in Water Treatment Technology Development. Monograph of Environmental Engineering Committee of Polish Academy of Sciences, 42, Lublin.

St.Clair J., Stamopoulos C. and Edwards M. (2012). Technical note: increased distance between galvanic lead: copper pipe connections decreases lead release. *Corrosion*, **68**(9), 779–783.

Stolzenburg T. R. and Nichols D. G. (1986). Effects of Filtration Method and Sampling on Inorganic Chemistry of Sampled Well Water, Proceedings of the Sixth National Symposium and Exposition on Aquifer Restoration and Ground Water Monitoring, National Water Well Association, Dublin, OH, pp. 216–234.

Stumm W. (1986). Investigations on the Corrosive Behavior of Waters. *A.S.C.E.*, **86**, 2657.

Świderska-Bróż M. (2001). Deterioration of water quality during treatment and distribution. *Inżynieria i Technika Środowiska*, **4**(3–4), (In Polish).

Szuster A. (2003). Secondary contamination of water in distribution network – problem to be solved. Przegląd Komunalny 1/2003,Wodociągi i Kanalizacja 2, (In Polish).

Szuster-Janiaczyk A. (2004). Water quality in distribution network and it's improvement after phosphate corrosion inhibitors dosing. *Przegląd Komunalny* **5**(152), (In Polish).

Szuster-Janiaczyk A. (2005). Processes occurring in water distribution systems – reasons, effects prevention. *Forum Eksploatatora*, **3**(20), (In Polish).

Szuster-Janiaczyk A. (2011). Methodology and selected results of implementadion of phosphate corrosion inhibitors in water in complex distribution systems. Doctoral Thesis (In Polish).

Tabesh M. and Saber H. (2012). A prioritization model for rehabilitation of water distribution networks using GIS. *Water Resources Management*, **26**, 225–241.

Tang Z., Hong S., Weizhong Xiao W. and Taylor J. (2006). Characteristics of iron corrosion scales established under blending of ground, surface, and saline waters and their impacts on iron release in the pipe distribution system. *Corrosion Science* **48**, 322–342.

Twort A. C., Ratnayaka D. D. and Brandt M. J. (2000). Water Supply. Arnold, London.

United Nations Economic Commission for Europe/WHO. (2007). Protocol on Water and Health. GE.06–26870-January 2007–4.290. Geneva.

US EPA (1982). Corrosion in potable water systems. Final report. U.S. Environmental Protection Agency, Washington D.C.

US EPA (1984). Corrosion Manual for Internal Corrosion of Water Distribution Systems. EPA 570/9-84-001. Office of Drinking Water. U.S. Environmental Protection Agency, Washington D.C.

US EPA (1994). Technical Notes on Drinking Water Methods. EPA/600/R-94/173. Office of Research and Develppment. Washington D.C.

US EPA (1996). Exposure Factors Handbook, vol. I – General Factors. U.S. Environmental Protection Agency, Washington, DC.

US EPA (1997). Manganese. Integrated Risk Information System. (IRIS). U.S. Environmental Protection Agency. Washington, DC.

US EPA (2004). Drinking Water Health Advisory for Manganese. United States Environmental Protection Agency, Office of Water, EPA-822-R-04-003, Washington, D.C.

US EPA (2009). White Paper on Rehabilitation of Wastewater Collection and Water Distribution Systems. United States Environmental Protection Agency, Office off Research and Development, Washington, DC. EPA/600/R-09/048.

US EPA (2011). Costs of arsenic removal technologies for small water systems: US EPA arsenic removal technology demonstration program; EPA/600/R-11/090.

Van den Hoven T. J. L., Buijs P. J., Jackson P. J., Gardner M., Leroy P., Baron J., Boireau A., Cordonnier J., Wagner I., do Mone H. M., Benoliel M. J., Papadopoulos I. and Quevauviller P. (1999). Developing a new protocol for the monitoring of lead in drinking water. EUR 19087.

Van Dijk J. C. and van der Kooij D. (2005). Water Quality 21 research programme for water supplies in The Netherlands. Water Science and Technology. *Water Supply*, **4**(5–6), 181–188.

Van Staden J. F., Mulaudzi L. V. and Stefan R. I. (2003). Speciation of Mn(II) and Mn(VII) by on-line spectrophotometric sequential injection analysis. *Analytica Chimica Acta* **499**, 129–137.

Ventura F., Matia L., Romero J., Boleda M. R., Marti I. and Martin J. (1995). Taste and odor events in Barcelona's water supply. *Wat. Sci. Tech.*, **31**(11), 63–68.

Vigeh M., Yokoyama K., Ramezanzadeh F., Dahaghin M., Fakhriazad E., Seyedaghamiri Z. and Araki S. (2008). Blood manganese concentrations and intrauterine growth restriction. *Reprod. Toxicol.*, **25**, 219–223.

Volk C., Dundore E., Schiermann J. and Lechevallier M. (2000). Practical evaluation of iron corrosion control in a drinking water distribution system. *Wat. Res.*, **34**(6), 1967–1974.

Vreeburg J. (2010). Discolouration in Drinking Water Systems. IWA Publishing, London.

Vreeburg J. H. G. (2007). Discolouration in drinking water systems: a particular approach. Ph.D. thesis, Civil Engineer, Delft.

Vreeburg J. H. G. and Boxall J. B. (2007). Discolouration in potable water distribution systems: a review. *Wat. Res.*, **41**, 519–529.

Vreeburg J. H. G., Schippers D., Verberk J. Q. J. C. and van Dijk J. C. (2008). Impact of particles on sediment accumulation in a drinking water distribution system. *Wat. Res.*, **42**, 4233–4242.

Wasserman G. A., Liu X., Parvez F., Ahsan H., Levy D., Factor-Litvak P., Kline J., van Geen A., Slavkovich V., LoIacono N. J., Cheng Z., Zheng Y. and Graziano J. H. (2006). Water manganese exposure and children's intellectual function in Araihazar, Bangladesh. *Environ Health Perspect.*, **114**(1), 124–129.

WHO (1958). International Standards for Drinking Water. World Health Organization, Geneva, Switzerland.

WHO (1984). Guidelines for Drinking-water Quality. 1st edn, World Health Organization, Geneva, Switzerland.

WHO (2002). Environmental Health Criteria 228, Principles and Methods for the Assessment of Risk from Essential Trace Elements, World Health Organization, Geneva, Switzerland.

WHO (2003). Iron in Drinking Water – Background Document for Development of WHO Guidelines for Drinking-Water Quality. World Health Organization, Geneva, Switzerland.

WHO (2004a). Guidelines for Drinking-water Quality. 3rd edn, World Health Organization, Geneva, Switzerland.

WHO (2004b). Manganese in drinking water – background document for development of WHO guidelines for drinking-water quality, WHO/SDE/WSH/03.04/104.

WHO (2006). Guidelines for Drinking-water Quality [electronic resource]: Incorporating first Addendum. Available: http://whqlibdoc.who.int/publications/2006/9241546964_eng.pdf.

WHO (2011a). Guidelines for Drinking-water Quality, 4th edn, World Health Organization, Geneva, Switzerland.

WHO (2011b). Manganese in Drinking-water Background document for development of WHO Guidelines for Drinking-water Quality, WHO/SDE/WSH/03.04/104/Rev/1.

Wilczak A., Jacangelo J. G., Marcinko J. P., Odell L. H., Kirmeyer G. J. and Wolfe R. L. (1996). Occurrence of nitrification in chloraminated distribution systems. *J. Am. Water Works Assoc.*, **88**(7), 74–85.

Williams S. M., Ainsworth R. G. and Elvidgc A. E. (1984). A Method of Assessing the Corrosivity of Waters Towards Iron. Water Research Centre Publication, Swindon, UK. 1984.

Wood R. J. (2009). Manganese and birth outcome. *Nutrition Reviews*, **67**, 416–420.

Woolf A., Wright R., Amarasiriwardena C. and Bellinger D. (2002). A child with chronic manganese exposure from drinking water. *Environmental Health Perspectives*, **110**(6), 613–116.

Wróbel G., Szymiczek M. and Wierzbicki L. (2004). Swaglining as a method of pipelines rehabilitation. *Journal of Material Processing Technology*, **157–158**, 637–642.

Young W. F., Horth H., Crane R., Ogden T. and Arnott M. (1996). Taste and odour thteshold concentrations of potential potable water contaminants. *Wat. Res.*, **30**(2), 331–340.

Zhang H. and Andrews S. A. (2012). Effects of phosphate-based corrosion inhibitors on the kinetics of chlorine degradation and HAA formation in contact with three metal materials. *Canadian Journal of Civil Engineering*, **39**, 44–54.

Zhang Y. and Edwards M. (2007). Anticipating effects of water quality changes on iron corrosion and red water. *Jour. Water Supply: Res & Technol.—AQUA*, **56**(1), 55.

Zimoch I., Koba B. and Trybulec K. (2004). GPW activities – water quality monitoring from source to the consumers tap. VI International Conference: Water Supply and Water Quality, Poznan. (In Polish).

Zota A. R., Ettinger A. S., Bouchard M., Amarasiriwardena C. J., Schwartz J., Hu H. and Wright R. O. (2009). Maternal blood manganese levels and infant birth weight. *Epidemiology*, **20**(3), 367–373.

IWA Publishing's authorised EU representative for General Product Safety Regulations is Diane D'Arras, 15 rue Duret, 75116 Paris, France, e-mail: safety@iwap.co.uk.

Printed and bound by CPI Group (UK) Ltd, Croydon, CR0 4YY
13/05/2026
02109623-0010